FEED THE BEES

Provide a Healthy Habitat to Help Pollinators Thrive

THE XERCES SOCIETY
FOR INVERTEBRATE CONSERVATION

Eric Lee-Mäder, Jarrod Fowler, Jillian Vento & Jennifer Hopwood

Storey Publishing

*The mission of Storey Publishing is to serve our customers by
publishing practical information that encourages
personal independence in harmony with the environment.*

Edited by Deborah Burns
Art direction and book design by Michaela Jebb
Text production by Erin Dawson

Cover photography by © Bryan Reynolds, back; © Saxon Holt, center front, spine, and back inset
All other front cover images: JP = © Jerry Pavia, SH = © Saxon Holt, BR = © Bryan Reynolds (rows left to right)
Row 1: © FLPA/Alamy, SH, JP, SH, © Mindy Fawver/Alamy, JP, SH, BR, JP, JP; Row 2: JP, BR, BR,
SH, © Steffan Hauser/botanikfoto/Alamy, JP, BR, JP, JP, JP; Row 3: BR, SH, JP, SH, SH, JP, BR, JP,
SH, BR; Row 4: JP, BR, JP, SH, JP, © Richard Becker/Alamy, JP, JP, © Premium Stock Photography
GmbH/Alamy, JP; Row 5: BR, SH, SH, SH, JP, SH, JP, JP, JP, BR, JP, SH; Row 6: JP, SH, JP, JP, SH,
SH, SH, SH, JP, BR, SH, JP; Row 7: SH, SH, SH, JP, JP, SH, SH, JP, JP, BR, SH, JP; Row 8: SH, JP,
BR, SH, BR, SH, SH, SH, JP, JP, JP, BR; Row 9: JP, SH, SH, SH, SH, SH, SH, BR, SH, SH, SH, SH;
Row 10: SH, BR, SH, SH, SH, SH, SH, SH, SH, SH, SH, JP

Interior photography credits appear on page 240
Illustrations by © David Wysotski, Allure Illustration, 11, 13, and bee silhouettes throughout
Range maps by Ilona Sherratt

© 2016 by The Xerces Society

Storey Publishing
210 MASS MoCA Way
North Adams, MA 01247
storey.com

Printed in China by Toppan Leefung Printing Ltd.
10 9 8 7 6 5 4 3 2 1

Library of Congress Cataloging-in-Publication Data on file

THIS BOOK IS DEDICATED TO EVERYONE who tears up their front yard to plant big chaotic wildflower gardens, to farmers who think hedgerows and wildflower field borders are just as important as crops, to urban planners and landscapers who turn gray and lifeless concrete landscapes into corridors of biodiversity, and to members of the Xerces Society.

Contents

What's Old Is New

DR. EDITH PATCH was the original insect conservationist, one of the first critics of indiscriminate pesticide use, an author of fantastically interesting children's books, and an early pioneer for women in science. But it was in her role as the first female president of the Entomological Society of America, at the organization's annual meeting in 1936, that Edith foreshadowed this exact book and everything we do here at the Xerces Society in a lecture titled "Without Benefit of Insects."

In that talk Edith discussed the wholesale destruction of insect life that had resulted from the new insecticide products that were being brought to market, and commented on how in attempting to control pests, we were destroying bees and beneficial insects. She challenged the assembled scientists to imagine a very different world in the year 2000, when she predicted that the president of the United States would issue a proclamation declaring that land areas at regular intervals throughout the country would be maintained as "Insect Gardens," directed by government entomologists. These would be planted with milkweed and other plants that could sustain populations of butterflies and bees. She then predicted that at some time in the future, *"Entomologists will be as much or more concerned with the conservation and preservation of beneficial insect life as they are now with the destruction of injurious insects."*

Although the exact year she predicted turned out to be slightly early, Edith was ultimately right. In the summer of 2015, after extensive behind-the-scenes talks between the White House, Xerces, and other conservation groups, President Barack Obama did indeed release a first-of-its-kind memorandum. It called upon all federal agencies to do two things:

1. To develop comprehensive conservation plans that would protect and restore habitat for bees and butterflies at federal facilities and on federal lands

2. To offer financial incentives for the restoration of pollinator habitat on private lands, especially farmlands

The president also issued a challenge to the conservation community to help foster a million new pollinator gardens in residential yards and business campuses across the country — an effort Xerces is supporting through our Bring Back the Pollinators Garden Campaign (xerces.org/bringbackthepollinators).

While the accuracy of Edith's prediction is both haunting and heartening, amazingly she was not alone in pioneering the call to create habitat for pollinators. Twenty years earlier, in fact, Iowa polymath Frank Chapman Pellett established near his childhood home what may have been the first large-scale bee garden in the United States. Although formally trained as a lawyer, Frank eschewed the basic trappings of prosperity, choosing instead to live in Gandhi-like midwestern simplicity in a small, plain farmhouse. There he researched tomato gardening, devoted countless hours to bird watching, and meticulously documented and cultivated the preferred wild pollen and nectar sources of his honey bees.

His ceaseless hours of observation resulted in the 1920 book *American Honey Plants*, possibly still the best publication of its kind in existence. In another book, *Our Backyard Neighbors*, Frank wrote of himself and his pollinator garden in the third person, saying, *"There were many wild flowers, such as asters and goldenrod, crownbeard and rudbeckia, which the neighbors regarded as weeds, but which the Naturalist guarded with jealous care."*

Along with Edith Patch and Frank Pellett, the late Canadian scientist Dr. Eva Crane played one of the largest roles in further inspiring early thinking about pollinator gardens. Although she was formally educated as a quantum mathematician and nuclear physicist, the gift of a beehive in 1942 (as a supplement to wartime sugar rationing) led Eva to devote the next five decades of her life to publishing nearly 200 books and articles on honey plants and indigenous beekeeping.

Her writing was based on her field research in more than 60 countries, where she often lived under primitive conditions, even in her later years. Her rigorous and exacting books, such as the *Directory of Important World Honey Sources,* are the most comprehensive attempts of their kind to document the nutritional value of pollen and nectar from thousands of species of plants, as well as those plants' potential honey yields.

In their own ways each of these deeply inquisitive champions of pollinator habitat inspired small communities of beekeepers and conservationists to see the landscape through a different lens. Plants previously scorned as weeds and unproductive "waste" areas on farms began to have value to at least a small segment of people, even as urbanization and agriculture intensified with enthusiasm.

By 1950 even the USDA Soil Conservation Service, the agency most responsible for saving American agriculture from itself during the Dust Bowl, recognized the value of pollinators. It distributed a simple educational bulletin to Midwest farmers, featuring an illustration of bumble bees flying between a hedgerow and a clover crop with the earnest title *Wild Bees Are Good Pollinators.* The bulletin lists important habitat areas on the farm to protect for pollinators, including stream banks, woodlots, shelterbelts, and field borders. For good effect the bulletin even features an illustration of a bag of clover seed with a dollar sign across its front.

Surprisingly, the dawn of the environmental movement brought little attention to pollinators during the 1960s, '70s, and '80s, although countless other important conservation issues finally received some long-overdue attention. Only when large-scale honey bee losses began to make headlines in 2006 did the conservation community again focus much on the role of pollinators. By that time several bumble bee species in the United States were dwindling toward extinction, and once-common monarch butterfly populations were in free fall. Now books and articles about pollinator conservation are everywhere. For those of us at Xerces who have been working on and writing about pollinators for decades, this long-overdue attention is gratifying and energizing.

The spiritual tradition of this particular book descends from Patch, Pellett, and Crane, but also from John Muir, Aldo Leopold, Rachel Carson, and many others. These are the writers who inspired us here at Xerces in our youth and early in our careers, and who ultimately helped bring us all together as the big extended family that we are today. Our goal, like that of the conservation writers who preceded us, is not just to preach the gospel, but also to invite you into the tribe. We hope that you will join us.

The initiation is simple: just plant flowers.

Plants and Pollinators: An Overview

WHEN WE OBSERVE ANIMALS pollinating nearly 90 percent of the plant species found on earth, we are witnessing a process more than 250 million years in the making. Sexual reproduction among plants, from a botanical standpoint, is nothing more than the transfer of pollen grains from a flower's male anthers to a flower's female stigmas, enabling fertilization. Once transferred, pollen grains germinate, grow pollen tubes into the plant's ovaries, and deliver gametes to produce seed and endosperm.

In very primitive plants, this process was carried out by wind or water. Between 245 million and 200 million years ago, however, the first flowering plants arose, with the earliest fossil records containing relatives of today's magnolias and water lilies. During this prehistoric time frame, flowering plants evolved two major reproductive adaptations: exposed male stamens that bear small, nutrient-rich pollen grains; and enclosed female carpels that protect ovules. These adaptations accelerated plant reproduction (and pollinator diversity), leading to diverse and dominant communities of flowering plants that almost 100 million years ago had spread across the globe.

ANATOMY OF A FLOWER

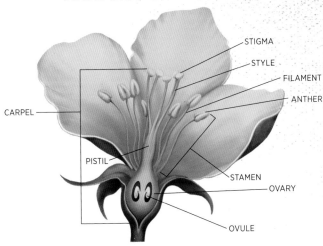

STIGMA

STYLE

FILAMENT

ANTHER

CARPEL

PISTIL

STAMEN

OVARY

OVULE

Plants Meet Pollinators

BEETLES, FLIES, AND WASPS are thought to be the first pollinators, accidentally spreading pollen while feeding on flowers. This set the stage for more complex plant-pollinator relationships to evolve, including prehistoric flowering plants that first attracted passive pollinators by providing sugary nectar, protein-packed pollen, fragrant resins, and vitamin-rich fats.

Flowers then responded to particular pollinators, coevolving with them to provide diverse bloom times, colors, scents, shapes, sizes, and rewards, and improving their reproductive efficiency. For example, flattened, large, scented, off-white flowers with accessible pollen, such as magnolia, attracted beetles, while tubular, large, scented, white flowers that bloom at night attracted moths.

Meanwhile, flowers also developed a variety of strategies to avoid self-fertilization and encourage genetic diversity:

- Self-incompatibility

- Physical distance between (male) anthers and (female) stigmas

- Male and female flower structures that are fertile at different times

- Separate male and female plants

Enter the Bees

The widespread distribution of diverse flowering plants 100 million years ago coincided with the appearance of intentional pollinators: bees. Bees are believed to have coevolved with flowers from predatory wasps. In general, both bees and wasps consume sugars as adults and proteins as larvae. Herbivorous bee larvae eat pollen as their protein source, however, while wasp larvae are typically carnivorous.

Pollen is essential for the reproduction of both bees and flowers, so the two groups have coevolved for mutual success. Adult bees evolved behavioral and physiological adaptations to gather and transport pollen more efficiently, such as:

BUZZ-POLLINATION. Flight muscles can create sound vibrations that dislodge pollen from flowers.

FLORAL CONSTANCY. An individual pollinator may specialize in foraging one flower type.

POLLEN-COLLECTING HAIRS. The "pollen basket" and other specialized hairs on a bee's body carry pollen back to the colony.

Although most bees are pollen generalists, capable of foraging on many plant species, many are specialists that forage on only a small group of specific flowers.

ANATOMY OF A HONEY BEE

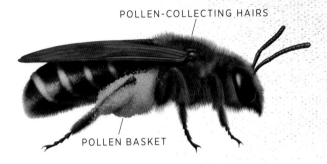

POLLEN-COLLECTING HAIRS

POLLEN BASKET

What Makes a Good Pollinator Plant?

A flower's color, odor, shape, size, timing, and reward (nectar or pollen) can increase or decrease the number of visits by specific pollinators. Some examples of how plants "reach out" to bees and others:

ULTRAVIOLET INVITATIONS. Bees can see ultraviolet light but not red light; thus, flowers in the ultraviolet range attract more bee visits, while red-hued flowers reduce them.

COLOR PHASES. Many flowers signal pollinators by changing color at different stages of development, attracting pollinators when they need them most, thus increasing the efficiency of the pollinators they depend upon.

NECTAR GUIDES. Contrasting patterns of flower shades, tints, and tones further direct pollinators toward floral rewards such as nectar or pollen, much like the nighttime runway lights of an airport.

FRAGRANCE. Minty or sweet, musky or ethereal, pungent or putrid, floral odors result from variations in chemical compounds. Fragrance can attract particular pollinators over long distances, varying in concentration and intensity according to species, flower age, and site conditions.

WITH ITS CONTRASTING COLORS,
this blanketflower ushers pollinators
toward the nectar and pollen at the
center of the bloom

Flower shape, size, and timing work together with color and odor to regulate pollinator visits. Abundant and diverse shapes and sizes, symmetrical or asymmetrical forms, arrangements on stems or branches in simple or complex groups, maturing at different rates: these variations can make it easier or harder for visitors to reach nectar and pollen.

For example, shallow, clustered flowers with landing platforms (such as sunflowers) have easily accessible floral rewards and attract many short-tongued pollinators such as sweat bees, beetles, and flies. In contrast, deep or tubular flowers without landing platforms often have hidden floral rewards accessible only by long-tongued or strong pollinators. A classic example of this latter flower type is bottle or closed gentian (*Gentiana* spp.), whose flowers remain closed and depend for pollination on bumble bees, which pry the petals apart and climb right inside.

Finally, many plants bloom according to a distinct seasonal rhythm — their **phenology** — which may be closely timed with the life cycle of specific pollinators. Others, meanwhile, bloom continuously or irregularly during the growing season, attracting many different types of pollinators. These rhythms can invite or exclude different pollinators depending upon the season or even the hour.

Risks and Rewards of Flower Foraging

OF COURSE, POLLINATORS MOST OFTEN visit flowers for nutrient-rich food rewards: pollen and nectar. The availability and quality of these rewards vary depending on time of day, environmental factors, and an individual plant's life cycle. And from the perspective of a bee, butterfly, or other pollinator, several factors can make a particular flower useful, or not.

Nutrition

Floral rewards include pollen, nectar, oils, and/or resins, depending on the plant species.

POLLEN, the most protein-rich of these rewards, is essential to bee reproduction. Once gathered, adult bees typically mix pollen with nectar and glandular secretions to form a nutritious "bee bread," which forms the diet of larval bees. Pollen grains vary from 10 to 100 micrometers in size, have distinctive shapes, and commonly contain protein levels ranging from 2 to 60 percent (including 10 essential amino acids, as well as varying concentrations of carbohydrates, lipids, sterols, and other micronutrients). While some bees, such as the common European honey bee, are generalist pollinators whose diets are not restricted to particular pollen types, others are specialists of pollen from particular flowers, including various mining bees, cellophane bees, and resin bees.

NECTAR is composed chiefly of carbohydrates and water, with low levels of amino acids, lipids, proteins, and various vitamins and minerals. Carbohydrates, primarily the sugars sucrose, fructose, and glucose, can range in concentrations from 10 to 70 percent based on species and weather. It is this sugar-rich food source that fuels adult bees, butterflies, and a myriad of other flower visitors, such as bats and hummingbirds. Nectar secretion, even within the same species of plant, can vary depending on humidity, precipitation, time of day, temperature, wind, latitude, soil, and various other factors. In turn, the pollinators visiting those blossoms may encounter short-term booms and busts of nectar availability.

OILS AND RESINS are secreted by some flowers to attract bees. Specialized floral glands produce calorie-rich, medicinal oils that are regularly collected by a few bees (for example, *Macropis* spp. and *Melitta* spp.) and mixed with pollen and nectar for feeding and medicating larvae. Most likely, such flower resins first evolved to protect the plants from herbivores or disease. Eventually bees came to use them as a food source, and as a resin for constructing antimicrobial and waterproof nests.

Nonfloral Rewards

Nonfloral (or "extrafloral") rewards include nectar, honeydew, fruits, and saps.

EXTRAFLORAL NECTAR is produced by many plants as sugary droplets from glands on leaves, stems, and other nonflowering plant parts. These nectar droplets attract beneficial predatory insects, such as ants, beetles, flies, mites, spiders, and wasps — all of which may attack plant pests. Among some plants, these extrafloral nectaries may supply even more nectar than the flowers do themselves. While less showy and aromatic than flowers, extrafloral nectaries are usually open and exposed for easy access by many types of beneficial insects (although not infrequently they are guarded by territorial ants!).

HONEYDEW is the sugary excrement of sap-feeding aphids, scale insects, whiteflies, and some butterfly caterpillars (mostly the blues, in the family Lycaenidae). Like extrafloral nectar, it is eagerly collected by many beneficial insects, including ants, bees, and wasps. In some locations, in fact, aphid honeydew is found in large enough quantities to produce small surplus honey crops. Honeydew is readily accessible but occasionally it, or the insects producing it, are guarded by ants. Think of ants tending aphids as though they were livestock, and you have a fairly accurate picture of this unique insect relationship.

PROPOLIS, also known as **bee glue**, is a resinous sap mixture collected from plants by bees and harvested by humans. Particular plants, including conifers and poplars, exude these resins from buds or from injuries as a natural antimicrobial defense.

Honey bees collect propolis to construct and defend hives, weatherproof small cracks and holes, smooth surfaces, dampen vibrations, and protect themselves from bacteria, fungi, mites, and other intruders. Humans harvest and use honey bee propolis in cosmetics, soaps, medicines, and wood polishes or varnishes.

Species of solitary mason bees also collect propolis to construct, partition, and seal nests.

Other Rewards

Beyond pollen and nectar, plants sustain pollinators in several other ways, and the most familiar of these is as caterpillar food for butterflies. With only a few exceptions, the vast majority of butterfly and moth caterpillars are herbivores that feed exclusively on plant foliage. Depending on the species, those caterpillars may be generalists, which can feed on many types of plants, or specialists with a very narrow range of plants on which they can successfully feed.

The specialists often acquire defensive chemical compounds from the plants they feed upon (such as alkaloids, cardenolides, or glycosides) that make those insects unpalatable or toxic to predators. For example, milkweed butterfly caterpillars such as the monarch and queen feed exclusively on milkweed (*Asclepias* spp.) foliage, which contains toxic cardenolides that repel most vertebrate predators.

Other than food resources, plants also offer nesting, egg-laying, and overwintering resources for pollinators, such as hollow or pithy canes; stalks, stems, or twigs; leaves, petals, or plant fibers; and exfoliating or peeling bark. Plants with hollow or pithy branches, such as brambles (*Rubus* spp.), elderberry (*Sambucus* spp.), and sumac (*Rhus* spp.), are used extensively as nesting spaces for a wide range of wild solitary bees and wasps.

Nearly 30 percent of North American native bee species nest in hollow stems or abandoned beetle borer holes — including leafcutter bees (*Megachile* spp.), mason bees (*Hoplitis* spp.; *Osmia* spp.), small carpenter bees (*Ceratina* spp.), and masked bees (*Hylaeus* spp.).

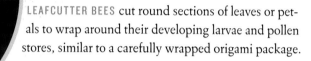

LEAFCUTTER BEES cut round sections of leaves or petals to wrap around their developing larvae and pollen stores, similar to a carefully wrapped origami package.

WOOL CARDER BEES (*Anthidium* spp.) comb plant fibers from the surface of fuzzy leaves and use them to create a wooly, felted plug that closes off the entrance to their nests inside hollow stems.

ABOVE Carpenter bee

GRASS-CARRYING WASPS (*Isodontia* spp.) gather grasses to plug up the entrance to their nests, building a grass barrier against other insects that would otherwise steal the food intended for their developing brood.

Risk Management

Foraging for food can be risky for pollinators. In the process of visiting flowers, an individual insect may encounter predators, disease vectors, or bad weather. The farther an insect has to travel, and the more energy it has to exert in collecting food, the more risk it is exposed to. Plants that provide an abundance of quickly accessible, nutrient-packed pollen and nectar obviously provide the greatest reward, and allow insect visitors to get on with the business of mating and reproduction.

Diversity in Time and Space

Landscapes with a wide diversity of blooms more effectively sustain pollinators throughout the seasons than do landscapes dominated by only a small handful of flowering plants. At a landscape scale, the presence or absence of different types of blooming plants can result in a "feast or famine" situation for pollinators. Thus, expansive landscapes of weedy or invasive plants such as purple loosestrife (*Lythrum salicaria*) or Himalayan blackberry (*Rubus armeniacus*) may provide an abundance of food for bees and other pollinators during their bloom period. Once the bloom is over, however, pollinators may suffer as those same invasive plants that temporarily sustained them now crowd out other types of wild plants that would otherwise have provided a variety of flower types throughout the entire growing season.

The Landscape around You

THIS VOLUME PRESENTS 100 useful pollinator plants that foster both generalist and specialist bees. The book is not exhaustive. All selections are adapted to North American climates with an emphasis on regionally native species. Where nonnative plants are included, we have focused on those species that are typically not invasive in most landscapes (although a few of the more tenacious nonnatives, such as sweetclover, deserve a bit of responsible scrutiny before introducing them to new places). Where the pollinator plants featured in this book are already present, they should be conserved — especially if they are unique native species.

Selecting Pollinator Plants

If your area has relatively few flowering plants, or if you wish to augment local plant populations, then this book will help guide you in that process. We have tried to provide a general overview of the preferred growing conditions and geographic range for each of the plants included in this book, with options for most of the United States and Canada. Of course, when selecting the plants featured in this book, it is also imperative to purchase only nursery plants that you know were not treated with long-lasting insecticides.

Creating Habitat

There are other fine books on habitat design for pollinators and beneficial insects, including the Xerces Society titles *Attracting Native Pollinators* and *Farming with Native Beneficial Insects*, so we do not cover that design and installation process in this book. However, here are some basic guidelines to consider when creating habitat for bees.

Provide large and contiguous habitat patches. Where possible, pollinator gardens, wildflower meadows, and habitat patches at least 5,000 square feet in size can offer a wonderfully productive land-scape feature for sustaining honey bees, butterflies, and countless wild bees alike. To sustain wild pollinators for crop production on farms, the current research suggests that 10 to 30 percent of a farm should be maintained in natural habitat to support both wild bees for crop pollination, and beneficial insects for natural pest control.

Within these areas, plants can be scattered about, but clumps or groupings of similar plants (of at least 4 square feet [1 square meter]) seem to be especially attractive to pollinators. This proxim-ity reduces their foraging time so that they can spend more time mating, nesting, and raising future generations of pollinators.

Plant diversity also enhances pollinator populations, as previ-ously mentioned. To attract a great diversity of wild bees, a land-scape should feature at least 12 to 20 species of flowering plants and have at least three species of blooming plants at any given time.

Most important, whatever you plant, the habitat must be pro-tected from insecticides (see page 22). We recommend at least a 50-foot-wide buffer (preferably 100-foot) between any pollinator habitat and areas such as cropland where insecticides are used.

Using Locally Native Plants

Native plants should always be prioritized in creating pollinator habitat. While nonnative species can provide complementary benefits (such as cover crop plants for enhancing soil health, or edible landscape plants such as fruit trees), native plants typically offer the best adaptation to their environment, and they have co-evolved with the many bees, butterflies, and other wildlife within their respective regions. Ideally, we encourage you to protect, collect, and sow seed from native plants that originate within or near your own community.

While native plants are ideal, introduced plants are often an irreversible presence in our humanized landscapes. Many of these species can offer copious floral rewards for pollinators. Select introduced plants with caution, however, and carefully avoid invasive or noxious plants to protect native plant communities and the wildlife that depend upon them.

Site Preparation

Finally, note that in landscapes heavily dominated by weeds and aggressive grasses, it can be hard to establish pollinator habitats such as prairies and meadows. Site preparation — removing those weeds before restoring an area in wildflowers — is challenging but essential. You can clear an area by cultivating, using prescribed fire, applying herbicides, and cover-cropping for one or more growing seasons.

At Xerces, our currently preferred method is soil solarization. This technique uses sheets of clear greenhouse plastic to cover, heat, and kill weeds and weed seeds over patches of ground for a full growing season, before removing the sheets and seeding new wildflowers into the clean area. We even save the plastic and reuse it over multiple seasons, slowly expanding habitat areas on farms and in gardens year after year. It works beautifully for us.

Once planted, these wildflower meadows tend to be long lived, although they do need occasional weeding and mowing to keep out invasive plants.

Pollinators and Pesticides

MAKE YOUR GARDEN A SAFE HAVEN FOR POLLINATORS BY AVOIDING PESTICIDE USE. Insecticides, including those used to control garden pests, tend to be broadly toxic to pollinators as well as to pests, and even some organic insecticides can pose a risk.

Pollinators are exposed to insecticides in gardens through direct contact with spray and residues on flowers or ingestion of insecticide residues present in pollen, nectar, or water. Pollinators can be killed outright by insecticidal exposure, but smaller, nonlethal doses can also be problematic. These lower doses can impair pollinators, altering their behavior, movement, growth, reproduction, and immune response to parasites and pathogens. Insecticide exposure can reduce pollinator populations quickly, and recovery to preexposure levels can require years.

Some plants for purchase at nurseries or home improvement stores have been treated with **neonicotinoid insecticides** during the production process. Neonicotinoids, a group of systemic chemicals, are long lasting in soil and within plants, and can be present in pollen and nectar of the flowers of treated plants long after you bring them home from the store. Even in small doses, neonicotinoids are harmful to pollinators.

You can prevent problems through careful garden management.

1. When purchasing plants, inquire at the nursery to be certain the ones you select are pesticide-free.

2. Monitor your plants to make sure they are healthy.

3. Plant a diversity of species to provide food and shelter for beneficial insects that will help control pest populations in your yard and garden.

Icon Key				
Honey Bee	Native Bee	Hummingbird	Butterfly	Moth

Native Wildflowers

Providing wildflower-rich habitat is the most significant action you can take to support pollinators. Native plants, which are adapted to local soils and climates, are usually the best sources of nectar and pollen for native pollinators. In addition, native plants often require less water than nonnatives, do not need fertilizers, and are less likely to become weedy. Here, we present some of our favorite North American native wildflowers for pollinators.

1.

ANISE HYSSOP, GIANT HYSSOP

(*Agastache* spp.)

AMONG THE MOST BEE-ATTRACTIVE PLANTS in their regions, more than a dozen species of giant hyssop are found in North America. Historically, mass plantings of one species, anise hyssop, were established in parts of the Midwest and Canada specifically as a honey plant. Nineteenth-century beekeeper accounts claimed that a single acre could provide ample forage for 100 colonies of bees, and that bees preferred giant hyssop even to sweetclover (see page 234). The sugar concentration in giant hyssop nectar reportedly exceeds 40%, and the resulting honey is light in color, slightly minty in flavor, and resistant to granulation.

EXPOSURE
Sun to part shade

SOIL MOISTURE
Average

RECOMMENDED SPECIES OR VARIETIES

Anise hyssop (*Agastache foeniculum*), purple giant hyssop (*A. scrophulariifolia*), and yellow giant hyssop (*A. nepetoides*), all widely distributed in the East and Midwest; nettle-leaf giant hyssop (*A. urticifolia*), found widely across the West. An introduced species, Korean mint (*A. rugosa*) adapts well to gardens in much of North America.

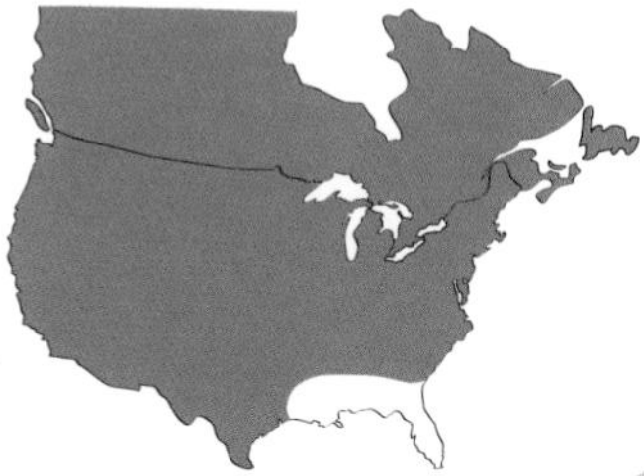

NOTABLE FLOWER VISITORS

Attracts a diverse variety of bees, butterflies, and occasionally hummingbirds. A small black sweat bee (*Dufourea monardae*) is a specialist of giant hyssop and beebalm in the Midwest.

ABOVE Monarch butterfly and beetles foraging giant hyssop

USES

Wildflower meadow/ prairie restoration

Farm buffer/filter strip

Ornamental

Edible/herbal/medicinal

BLOOM TIME
Summer

FLOWER COLOR
Purple, white

MAXIMUM HEIGHT
6+ feet (1.8+ m)

2. ASTER

(*Symphyotrichum* spp.)

AN IMPORTANT LATE-FALL FOOD SOURCE for bees, asters can help new bumble bee queens build up their energy reserves prior to winter dormancy. In some regions they are a late-season honey plant (although the table quality of the product is not well regarded). Various species are well adapted to upland or wetland conditions and to open sunny meadows or forest edges. The reported sugar concentration in the nectar of some aster species ranges from 24 to 41%.

EXPOSURE

Sun to part shade

SOIL MOISTURE

Wet to dry,
depending on species

LEFT Leafcutter bee foraging aster

RECOMMENDED SPECIES OR VARIETIES

For the East and Midwest, smooth blue aster (*Symphyotrichum laeve*), New England aster (*S. novae-angliae*), western silver aster (*S. sericeum*), New York aster (*S. novi-belgii*), and arrow-leaved aster (*S. sagittifolium*). In some Western locations, Pacific aster (*S. chilense*) and Douglas aster (*S. subspicatum*).

USES

Hedgerow

Wildflower meadow/ prairie restoration

Wetland restoration

Farm buffer/filter strip

Pollinator nesting material or caterpillar host plant

Ornamental

NOTABLE FLOWER VISITORS

Attracts bumble bees (*Bombus* spp.). Aster specialists include mining bees (*Andrena hirticincta, A. asteris, A. asteroides, A. nubecula, A. placata, A. simplex, A. solidaginis, Pseudopanurgus nebrascensis*), the polyester bee (*Colletes simulans armatus*), and the long-horned bee (*Melissodes druriella*). Host plant for caterpillars of the pearl crescent butterfly (*Phyciodes tharos*), the northern crescent (*P. selenis*), and the arcigera flower moth (*Schinia arcigera*). In the Midwest, various asters may also help fuel the fall monarch butterfly migration.

BLOOM TIME
Typically late summer through fall

FLOWER COLOR
Typically white, pink, blue, or purple

MAXIMUM HEIGHT
6+ feet (1.8+ m)

3.

BEEBALM

(Monarda spp.)

ALL BEEBALMS are excellent pollinator plants, although the types of pollinators they attract can range widely among species. Various species can be locally common. As a rule, beebalms establish quickly from seed when competition from other plants is minimal, and they are often among the first native plants to appear in newly seeded meadows and restored prairies. With care, beebalm usually flowers in abundance in such restoration plantings.

EXPOSURE
Sun to part shade

SOIL MOISTURE
Average to dry

RECOMMENDED SPECIES OR VARIETIES

In the Midwest and East, lavender-flowered wild bergamot (*Monarda fistulosa*) and similar red-flowered scarlet beebalm (*M. didyma*). In the South and Southwest, annual lemon beebalm (*M. citriodora*), which closely resembles spotted beebalm (*M. punctata*). Spotted beebalm is a short-lived perennial that tolerates drier and sandy soils; it is a top honey plant, with reported honey yields of up to 500 pounds per acre.

USES

Wildflower meadow/ prairie restoration

Farm buffer/filter strip

Pollinator nesting material or caterpillar host plant

Ornamental

Edible/herbal/medicinal

NOTABLE FLOWER VISITORS

Wild bergamot attracts bumble bees, hummingbirds, and hawk moths; spotted beebalm attracts a wide variety of bee species. A small black sweat bee, *Dufourea monardae*, is a specialist of beebalm in the Midwest and Northeast; others include *Perdita gerardiae* and *Protandrena abdominalis*. Various beebalms are host plants for caterpillars of raspberry pyrausta (*Pyrausta signatalis*), orange mint (*P. orphisalis*), and hermit sphinx (*Lintneria eremitus*) moths. Researchers in mid-Atlantic states have recently observed sand wasps (*Bicyrtes*) using spotted beebalm extensively for nectar. These wasps are voracious predators of brown marmorated stinkbug (*Halyomorpha halys*), a significant pest of orchards and vegetable crops.

BLOOM TIME	FLOWER COLOR	MAXIMUM HEIGHT
Midsummer	Lavender, red, purple, white, pink	4 feet (1.2 m)

4. BLACK-EYED SUSAN

(*Rudbeckia* spp.)

ALTHOUGH VALUABLE TO VARIOUS BUTTERFLIES, black-eyed Susan and its relatives tend to attract fewer numbers of bees. It is one of the easiest native wildflowers to establish from seed and worth including in wildflower seed mixes throughout its native range.

EXPOSURE
Full sun

SOIL MOISTURE
Dry to wet,
depending on species

RECOMMENDED SPECIES OR VARIETIES

Midwest and East: black-eyed Susan (*Rudbeckia hirta*), common and widely adapted, found from central Canada east to the Maritimes and south to the Gulf Coast and Florida; cutleaf coneflower (*R. lanciniata*), more typical of stream banks, wetland edges, and moist forest edges; brown-eyed Susan (*R. triloba*), a short-lived perennial that grows tall and bushy where soils are fertile, sites sunny, and competition absent. Western species are usually restricted to high alpine meadows and not commercially available.

NOTABLE FLOWER VISITORS

Andrena rudbeckiae is a specialist bee of black-eyed Susan. The plant attracts various long-horned bees (in the genus *Melissodes*) and also provides food for the caterpillars of bordered patch (*Chlosyne lacinia*), gorgone checkerspot (*C. gorgone*), and silvery checkerspot (*C. nycteis*) butterflies. Cutleaf coneflower attracts honey bees.

USES

Wildflower meadow/
prairie restoration

Wetland restoration

Reclaimed industrial
land/tough sites

Rangeland/pasture

Farm buffer/filter strip

Pollinator nesting material
or caterpillar host plant

Ornamental

LEFT Pollen-laden long-horned bee on black-eyed Susan

BLOOM TIME	FLOWER COLOR	MAXIMUM HEIGHT
Summer	Yellow	6+ feet (1.8+ m)

5.
BLANKETFLOWER
(*Gaillardia* spp.)

THESE BRIGHTLY COLORED annual and perennial wildflowers are both drought tolerant and long blooming, attracting many native bees as well as honey bees. They produce good-quality, dark amber honey; the average sugar concentration in the nectar has been reported at approximately 32%. Annual species tend to be easier to establish from seed than perennial species.

EXPOSURE
Sun

SOIL MOISTURE
Average to dry

RECOMMENDED SPECIES OR VARIETIES

Nearly a dozen species of blanketflower are found in North America, but only two are widely available as seed or garden plants. Blanketflower (*Gaillardia aristata*) is a perennial species of the northern plains, Rocky Mountains, and inland Northwest. Indian blanket (*G. pulchella*) is an annual species occurring from Arizona across the southern plains, Gulf Coast, and Florida.

NOTABLE FLOWER VISITORS

Attracts many wild bee species, including various leafcutter bees (*Megachile* spp.) and green metallic sweat bees (*Agapostemon* spp.). Provides food for caterpillars of the bordered patch butterfly (*Chlosyne lacinia*) and the brilliantly colored gaillardia flower moth (*Schinia masoni*) and painted schinia (*S. volupia*). These latter two butterflies have wing patterns and colors that mimic blanketflower's petals.

USES

Wildflower meadow/ prairie restoration

Rangeland/pasture

Farm buffer/filter strip

Pollinator nesting material or caterpillar host plant

Ornamental

LEFT Leafcutter bee

BLOOM TIME	FLOWER COLOR	MAXIMUM HEIGHT
Summer	Orange, yellow	2 feet (0.6 m)

6. BLAZING STAR
(*Liatris* spp.)

EXCELLENT NATIVE POLLINATOR PLANTS, blazing stars grow from a tuberlike corm and will often tolerate poor soil conditions. The spikes bloom from the top down for several weeks. Scientists at the Xerces Society and native seed nursery partners are investigating the role of chemical cues in attracting monarch butterflies to blazing star from considerable distances away.

EXPOSURE
Sun

SOIL MOISTURE
Dry to moist

ABOVE Southern plains bumble bee
(*Bombus fraternus*) foraging blazing star

RECOMMENDED SPECIES OR VARIETIES

Meadow blazing star (*Liatris ligulistylis*)
deserves special note as a monarch butterfly
magnet. The towering prairie blazing star (*L. pyc-nostachya*), the smaller cylindrical (*L. cylindrica*),
marsh (*L. spicata*), and rough (*L. aspera*) are all
excellent selections.

NOTABLE FLOWER VISITORS

Attracts long- and short-tongued bumble bees
and butterflies. Meadow blazing star (*Liatris
ligulistylis*) attracts monarch butterflies in strik-ing numbers. Caterpillar host plant for the pink-colored bleeding flower moth (*Schinia sanguinea*).

USES

Wildflower meadow/
prairie restoration

Farm buffer/filter strip

Pollinator nesting material
or caterpillar host plant

Ornamental

BLOOM TIME	FLOWER COLOR	MAXIMUM HEIGHT
Summer	Lavender	6 feet (1.8 m)

7. BLUE CURLS

(Trichostema lanceolatum)

BEFORE ITS NATIVE GRASSLANDS were plowed into farm fields, making it scarcer across the landscape, blue curls, also known as vinegar weed, was considered the most important late-season honey plant in California. Accounts from a century ago reported routine honey yields averaging 80 pounds per colony and described the honey as white and quick to granulate. The average sugar concentration in the nectar of blue curls has been reported at 27%. As an annual plant, blue curls probably requires some periodic disturbance, such as fire or occasional grazing, to encourage reseeding. As with other native species, however, too much disturbance or crowding by invasive species will reduce its numbers.

EXPOSURE
Sun

SOIL MOISTURE
Average to dry

RECOMMENDED SPECIES OR VARIETIES

Although a number of closely related species are found across the United States, only the native California blue curls (*Trichostema lanceolatum*) is documented as a significant bee plant. It is likely that the larger shrubby, evergreen wooly blue curls (*T. lanatum*) is also a valuable pollinator plant. This species is restricted to southern California.

NOTABLE FLOWER VISITORS

Attracts honey bees, which suggests that many other pollinators such as native bees and butterflies are likely to visit as well.

USES

Wildflower meadow/prairie restoration

Rangeland/pasture

LEFT Honey bee

BLOOM TIME	FLOWER COLOR	MAXIMUM HEIGHT
Fall	Blue	2 feet (0.6 m)

8. BLUE VERVAIN

(*Verbena hastata*)

VALUABLE FOR ATTRACTING a diversity of pollinators, blue vervain is able to establish quickly and compete with other vegetation. This wetland-adapted plant grows from fibrous root systems and rhizomes and spreads into small colonies. Because of this growth pattern, a casual observer might consider blue vervain weedy, but it is a native species. Seed for this plant can be relatively inexpensive, and its easy establishment in fertile and wet soil conditions makes it a good choice for wet prairies, pastures, roadside ditches, rain gardens, and similar sites.

EXPOSURE
Sun

SOIL MOISTURE
Average to wet

RECOMMENDED SPECIES OR VARIETIES

Blue vervain (*Verbena hastata*) occurs primarily in the Midwest, Northeast, and across eastern Canada, typically in lowlands with damp soil. The similar-looking hoary vervain (*V. stricta*) is found primarily in the Midwest in dry, upland soils and is a common pasture plant in some locations. Both are excellent pollinator plants.

USES

Wildflower meadow/ prairie restoration

Wetland restoration

Farm buffer/filter strip

Pollinator nesting material or caterpillar host plant

NOTABLE FLOWER VISITORS

Attracts bees; host plant for caterpillars of the common buckeye butterfly (*Junonia coenia*), the verbena moth (*Crambodes talidiformis*), and the verbena bud moth (*Endothenia hebesana*).

BLOOM TIME	FLOWER COLOR	MAXIMUM HEIGHT
Summer	Blue	6 feet (1.8 m)

9.
CALIFORNIA POPPY
(*Eschscholzia californica*)

THIS CHEERY, COLORFUL ANNUAL or short-lived perennial attracts a diversity of bee species. Self-sowing and generally drought tolerant, California poppy grows well in disturbed areas, following fires, and along roadsides and railroad rights-of-way. The petals close at night and open again the following morning, and they may also close during rainy or overcast weather.

Some West Coast botanists and ecologists recognize distinct subspecies or locally unique populations, although widespread ornamental plantings (as well as stowaway seeds in automobile tire treads and on shoes) have contributed to cross-breeding.

EXPOSURE
Sun to part shade

SOIL MOISTURE
Dry

RECOMMENDED SPECIES OR VARIETIES

Ornamental variants include cultivars with red, pink, or white petals or double flowers, but the common type, with orange to pale yellow flowers, seems to attract the most bees.

NOTABLE FLOWER VISITORS

Attracts bumble bees, especially the common yellow-faced bumble bee (*Bombus vosnesenskii*). Most visiting bees appear primarily to collect pollen, suggesting the flowers produce little nectar. Butterflies and hummingbirds ignore the plant entirely.

USES

Wildflower meadow/ prairie restoration

Reclaimed industrial land/tough sites

Rangeland/ pasture

Ornamental

BLOOM TIME
Late spring

FLOWER COLOR
Yellow, orange

MAXIMUM HEIGHT
1 foot (30 cm)

10. CLARKIA
(*Clarkia* spp.)

THESE EARLY BLOOMERS produce showy flowers that support certain pollinators when most blossoms are scarce. While the *Clarkia* genus includes more than 30 species, almost all are restricted to California and adjacent states, such as Oregon. Most are annual wildflowers, and many are not commercially available. A few species, including the common farewell to spring or godetia (*Clarkia amoena*), have made their way into the ornamental plant trade. Seed for these is widely available. They are probably not prolific nectar or pollen producers, but they have the distinct advantage of blooming in early to midsummer when most other West Coast wildflowers are finished for the year. This bloom time, along with the low cost of seed, fast establishment, and aggressive reseeding, makes them a valuable addition to pollinator meadows along the West Coast.

EXPOSURE
Sun

SOIL MOISTURE
Average

RECOMMENDED SPECIES OR VARIETIES

Farewell to spring (*C. amoena*) is the most common and widely available. Other species are increasingly available, including mountain garland (*C. unguiculata*) and winecup clarkia (*C. purpurea*).

USES

Wildflower meadow/ prairie restoration

Rangeland/pasture

Ornamental

NOTABLE FLOWER VISITORS

Mostly attracts small, drab-colored sweat bees. Butterflies, hummingbirds, and large bees tend to ignore clarkia, despite its large, showy flowers.

BLOOM TIME	FLOWER COLOR	MAXIMUM HEIGHT
Early summer	Pink	2 feet (0.6 m)

11. COREOPSIS
(*Coreopsis* spp.)

SHOWY SUMMER BLOOMERS, these wildflowers are not necessarily "pollinator magnets," but they are dependable workhorse plants. In prairies, meadows, and disturbed sites they attract a moderate diversity (if not always an abundance) of insects. Their low cost, easy establishment, and tolerance for dry, sandy sites where other plants struggle make them useful additions to pollinator seed mixes.

EXPOSURE
Sun

SOIL MOISTURE
Average to dry

RECOMMENDED SPECIES OR VARIETIES

Of the roughly two dozen perennial and annual species in the United States, lance-leaved coreopsis (*Coreopsis lanceolata*) is especially adaptable, widespread, and tough. It's a good choice for meadow plantings and semidisturbed areas such as roadsides and contour buffer strips on farms. The seed of annual plains coreopsis (*C. tinctoria*) is very inexpensive and easy to grow on bare soil under a variety of conditions. This species does not reseed well, however, or compete well with perennial vegetation.

NOTABLE FLOWER VISITORS

Primarily attracts small native sweat bees, native sunflower bees (*Svastra* spp.), long-horned bees (*Melissodes* spp.), and hover flies. Attracts relatively few butterflies, honey bees, or bumble bees.

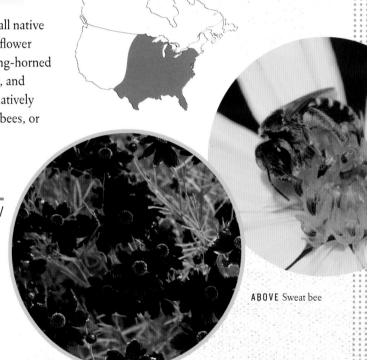

USES

Wildflower meadow/ prairie restoration

Rangeland/pasture

Farm buffer/filter strip

Ornamental

ABOVE Sweat bee

BLOOM TIME	FLOWER COLOR	MAXIMUM HEIGHT
Summer	Yellow, orange, red	2 feet (0.6 m)

12.

CULVER'S ROOT

(Veronicastrum virginicum)

EXPOSURE
Sun

SOIL MOISTURE
Average to wet

A LTHOUGH THE SEASON OF BLOOM is relatively short, Culver's root flowers are showy and extremely attractive to many bees and some butterflies. It's common to see honey bees mobbing the slender spikes when the plant is in flower, suggesting that it offers very high-quality nectar. As the plant does not occur in large, dense populations, it has not attracted significant attention from beekeepers. This perennial does best in rich, fertile, and moist soils and will not thrive in dry sites. In shady locations it topples over. It's an excellent plant for warm, open, wet prairies, meadows, or swamp edges.

RECOMMENDED SPECIES OR VARIETIES

L avender-flowered ornamental selections are now available, although the white-flowered type is still most likely the most attractive to bees.

NOTABLE FLOWER VISITORS

A ttracts honey bees, bumble bees, and various wild solitary bees.

USES

Wildflower meadow/ prairie restoration

Wetland restoration

Ornamental

BLOOM TIME
Summer

FLOWER COLOR
White, lavender

MAXIMUM HEIGHT
6 feet (1.8 m)

13.
CUP PLANT, COMPASS PLANT, ROSINWEED

(*Silphium* spp.)

THESE PRAIRIE PLANTS, tall and sunflower-like, have deep taproots, tough, papery leaves, and very long life spans. Once locally common, these enormous wildflowers have disappeared from the landscape with the loss of native prairies. In addition to attracting pollinators, the *Silphiums* provide excellent seed for songbirds. The common name of cup plant refers to the leaves that clasp the stem, forming cups that collect rainwater. It's common to see birds drinking from these cups.

EXPOSURE
Sun

SOIL MOISTURE
Moist to dry

RECOMMENDED SPECIES OR VARIETIES

Cup plant (*Silphium perfoliatum*), perhaps the most widely distributed species, will form small colonies in optimal conditions, with very fertile, moist, deep soils — picture a giant clonal colony growing next to your compost pile. Compass plant (*S. laciniatum*) has deeply lobed leaves that typically align along a north–south axis, giving the plant its common name. Starry rosinweed (*S. asteriscus*) is found throughout the Deep South, extending to the tip of Florida, and is a magnet for showy swallowtail butterflies and many other insects.

USES

Wildflower meadow/ prairie restoration

Rangeland/pasture

Pollinator nesting material or caterpillar host plant

Ornamental

NOTABLE FLOWER VISITORS

Attracts many visitors, including honey bees, bumble bees, and big showy butterflies. Soldier beetles and adult fireflies are unexpected but common visitors. One specialist is *Dieunomia heteropoda*, the largest sweat bee in the eastern United States, a striking species with black wings, a black body, long, curled antenna, and legs graced with unusual hooks and projections. Host plant for caterpillars of the silphium moth (*Tabenna silphiella*), and for the rare prairie cicada (*Okanagana balli*, not a pollinator), which spends most of its life burrowing and feeding within the thick woody stems and crowns of compass plant, much as other cicadas burrow and feed within trees. Leafcutter bees use broken, hollow stems of cup plant and its relatives as nest sites.

BLOOM TIME	FLOWER COLOR	MAXIMUM HEIGHT
Summer	Yellow	7 feet (2 m)

14. FIGWORT
(*Scrophularia* spp.)

PROLIFIC NECTAR PRODUCERS, figworts in a solid stand can leave visitors' clothes and skin wet with sticky nectar. One species, carpenter's square (*Scrophularia marilandica*), is also known as Simpson's honey plant. In the 1880s it was mass-planted in parts of the Midwest; beekeepers claimed a single acre could produce 400 to 500 pounds of honey. The flower nectar has been described as 18 to 32% sugar, and the resulting honey is clear, light yellow, and aroma-free. Because they tolerate partial shade and damp conditions, figworts are excellent for river bottom plantings and wetland edges.

EXPOSURE
Part shade

SOIL MOISTURE
Average to damp

RECOMMENDED SPECIES OR VARIETIES

Carpenter's square is limited to the eastern United States and Canada. More widely distributed is lance-leaved figwort (*Scrophularia lanceolata*), which occurs across most of North America outside of the Deep South. Several less common species are found in other regions. Because their flowers are not showy (although interesting up close), figworts are available only from specialist nurseries, but worth seeking out.

NOTABLE FLOWER VISITORS

Attracts huge numbers of bees, wasps, flies, and hummingbirds, especially when planted in large clusters.

USES

Hedgerow

Wildflower meadow/ prairie restoration

Reforestation/shade garden

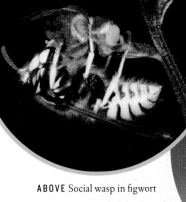

ABOVE Social wasp in figwort

BLOOM TIME	FLOWER COLOR	MAXIMUM HEIGHT
Late spring to summer	Green, red	6 feet (1.8 m)

15. FIREWEED

(Chamerion angustifolium)

A PLANT OF COOL CLIMATES and high altitudes, fireweed thrives in moist areas where fires have eliminated woody plants. Its seeds can remain viable in the soil for long periods, then rapidly germinate when forest clearings occur, such as after a fire — hence its common name. In such areas fireweed can suddenly appear in tremendous abundance but then quickly fade in subsequent years as the forest grows back, or as unfavorable weather patterns (such as drought) reduce flowering. Fireweed offers a long show, typically from June through August, with new blossoms opening on flower spikes to replace old blooms over many weeks.

Fireweed is not commonly available from seed, since the seed is tiny and attached to silky filaments, making planting a challenge. It grows easily from root or rhizome cuttings and will rapidly expand in large colonies as long as space and soil conditions are favorable. In the Pacific Northwest where fireweed reaches its optimal height, it's an

EXPOSURE
Sun to part shade

SOIL MOISTURE
Average

ideal hedgerow plant, mixing well with small shrubs such as Nootka rose. Fireweed is considered one of the most prolific honey plants in the Northern Hemisphere, with honey yields from 50 to 125 pounds per colony. The honey is light in color, sometimes described as "lightly spicy" or "buttery," and highly valued as a premium single-source honey. The average sugar concentration in the nectar of fireweed has been reported at 35%.

RECOMMENDED SPECIES OR VARIETIES

The common wild type is pink; white-flowered cultivars are available from specialty nurseries.

NOTABLE FLOWER VISITORS

Attracts honey bees, bumble bees, hummingbirds, and various solitary wild bees. Host plant for caterpillars of the bedstraw (*Hyles gallii*) and white-lined sphinx moths (*H. lineata*).

USES

Hedgerow

**Wildflower meadow/
prairie restoration**

**Pollinator nesting
material or caterpillar
host plant**

Ornamental

Edible/herbal/medicinal

BLOOM TIME	FLOWER COLOR	MAXIMUM HEIGHT
Summer	Pink	6 feet (2 m)

16. GLOBE GILIA

(Gilia capitata)

A WEST COAST ANNUAL, globe gilia grows quickly on warm, bare soil. Its seed is inexpensive and available in bulk quantities from many wildflower seed companies, facilitating large-scale plantings along the edges of farm fields or roadsides. Along the West Coast, the Xerces Society includes globe gilia with meadowfoam, California poppy, lupines, lacy phacelia, and other annuals in mass plantings adjacent to almond orchards and blueberry farms to attract and sustain mason bees, mining bees, and other wild pollinators.

EXPOSURE
Sun

SOIL MOISTURE
Average to dry

RECOMMENDED SPECIES OR VARIETIES

Closely related bird's eye gilia (*Gilia tricolor*) looks nothing like globe gilia but is nonetheless a good bee plant. It's available from some wild-flower seed vendors.

USES

Wildflower meadow/ prairie restoration

Rangeland/pasture

Ornamental

NOTABLE FLOWER VISITORS

Attracts mason bees, mining bees, syrphid flies, and small wasps.

LEFT Small white butterfly on globe gilia

BLOOM TIME	FLOWER COLOR	MAXIMUM HEIGHT
Spring to summer	Blue	2 feet (0.6 m)

17. GOLDENROD
(*Solidago* spp.)

GOLDENRODS ARE AMONG the most important late-season pollinator plants. Honey bees frequently collect large amounts of goldenrod nectar prior to winter; other bees use the pollen to provision late-season nests. Many beekeepers in the Northeast depend on goldenrod as their colonies' primary winter food source and often report honey gains of 50 to 80 pounds per colony. Goldenrod honey is typically dark, thick, somewhat pungent in aroma, and quick to granulate. The average sugar concentration in the nectar of some goldenrod species has been reported at approximately 33%.

The number of goldenrod species is vast, and it can be difficult to distinguish among them. While western states and provinces lack the sheer abundance of goldenrods found in the East, a locally adapted goldenrod probably exists anywhere you are. In general, goldenrods thrive in open areas with occasional mowing to remove competition from trees and shrubs.

EXPOSURE
Sun to part shade

SOIL MOISTURE
Average

Some of the easier-to-identify species such as showy goldenrod (*Solidago speciosa*), Riddell's goldenrod (*S. riddellii*), and stiff goldenrod (*S. rigida*) are easily available from native plant nurseries; all are excellent bee plants. Seaside goldenrod (*S. sempervirens*), a native of the Atlantic coast, blooms during the fall monarch butterfly migration and provides an important nectar source for the travelers.

USES

**Wildflower meadow/
prairie restoration**

**Reclaimed industrial
land/tough sites**

Rangeland/pasture

Farm buffer/filter strip

Ornamental

NOTABLE FLOWER VISITORS

Attracts many solitary wasps, fireflies, soldier beetles (especially *Chauliognathus pennsylvanicus*). Specialist bees include mining bees (*Andrena hirticincta, A. nubecula, A. placata, A. simplex,* and *A. solidaginis*), the polyester bee (*Colletes simulans armatus*), and the long-horned bee (*Melissodes druriella*). Other specialist bees include *Andrena asteris, A. canadensis, Perdita octomaculata,* and *Colletes solidaginis.*

BLOOM TIME	FLOWER COLOR	MAXIMUM HEIGHT
Late summer to fall	Yellow, white	6+ feet (2+ m)

18. GUMWEED
(*Grindelia* spp.)

TOUGH, RESINOUS, LEATHERY-LEAVED, gumweeds are sometimes viewed as weedy rangeland plants, getting little attention because they don't provide good fodder for livestock and their honey is considered inferior and granular. Nonetheless, in some areas of the West these perennials are among the best summer plants for wild bees. This is especially true along the West Coast, where most wildflowers bloom in spring immediately after the winter rains, leaving a shortage of blooming plants in the dry summer and fall. Tough, drought-tolerant gumweed fills this gap.

EXPOSURE
Sun

SOIL MOISTURE
Average

RECOMMENDED SPECIES OR VARIETIES

In the dry inland West and across California, curlycup gumweed (*Grindelia squarrosa*) is the most common and adaptable species. In the wet areas of the Pacific Northwest, Puget Sound gumweed (*G. integrifolia*) is better adapted, and one of the best summer plants for attracting native bees in the region. Unfortunately, the seed of these and other gumweeds is available from very few commercial suppliers.

NOTABLE FLOWER VISITORS

Attracts mostly summer bees such as various leafcutter bees (*Megachile* spp.), long-horned bees (*Melissodes* spp.), and green metallic sweat bees (*Agapostemon* spp.).

USES

Wildflower meadow/ prairie restoration

Rangeland/pasture

Edible/herbal/ medicinal

BLOOM TIME	FLOWER COLOR	MAXIMUM HEIGHT
Summer through late fall	Yellow	5 feet (1.5 m)

19. IRONWEED
(*Vernonia* spp.)

THE VERY SHOWY BLOSSOMS of these tall plants may be short lived, but they attract numerous types of bees and butterflies. More than a dozen native ironweed species are found across North America, although many have very limited ranges. Some are plants of dry, upland conditions, while others prefer wetland edges. Ironweeds are not typically regarded as honey plants.

EXPOSURE
Sun

SOIL MOISTURE
Average to wet

ABOVE Orange acraea butterfly foraging ironweed

RECOMMENDED SPECIES

Some of the more widely distributed and adaptable species include prairie ironweed (*Vernonia fasciculata*), New York ironweed (*V. noveboracensis*), giant ironweed (*V. gigantea*), and Missouri ironweed (*V. missurica*).

USES

Wildflower meadow

Wetland restoration

Pollinator nesting material or caterpillar host plant

Ornamental

NOTABLE FLOWER VISITORS

Several wild bees are specialist pollen collectors of ironweed, including the long-horned bees *Melissodes denticulata* and *M. vernoniae*. Host plant for caterpillars of the ironweed borer moth (*Papaipema cerussata*), the Parthenice tiger moth (*Grammia parthenice*), and the red groundling moth (*Perigea xanthioides*).

BLOOM TIME	FLOWER COLOR	MAXIMUM HEIGHT
Summer	Purple	7 feet (2 m)

20.
JOE-PYE WEED, BONESET

(*Eutrochium* spp., *Eupatorium perfoliatum*)

PRIMARILY KNOWN as butterfly plants, the Joe-Pye weeds and their close relative boneset also attract many solitary bees, bumble bees, and other insects. In general, these are plants of damp soils and sunny open areas, or the open edges of forests. Ditches and river bottoms are common natural locations for boneset and Joe-Pye weed.

EXPOSURE
Sun to part shade

SOIL MOISTURE
Average to wet

RECOMMENDED SPECIES OR VARIETIES

Several species of Joe-Pye weed are easily available from native plant nurseries and seed companies. Common, widely distributed options include spotted Joe-Pye weed (*Eutrochium maculatum*), hollow Joe-Pye weed (*E. fistulosum*), and sweetscented Joe-Pye weed (*E. purpureum*). Boneset (*Eupatorium perfoliatum*) is also available from native plant specialists and attracts an amazing diversity of insects.

USES

Wildflower meadow/ prairie restoration

Wetland restoration

Pollinator nesting material or caterpillar host plant

Ornamental

NOTABLE FLOWER VISITORS

Attracts big, showy butterflies such as monarchs and swallowtails, but also various solitary bees, bumble bees, and other insects. Boneset attracts many of those and typically more, including various beneficial predatory wasps and beetles. Host plant for caterpillars of the ruby tiger moth (*Phragmatobia fuliginosa*), the three-lined flower moth (*Schinia trifascia*), the boneset borer moth (*Carmenta pyralidiformis*), and the clymene moth (*Haploa clymene*).

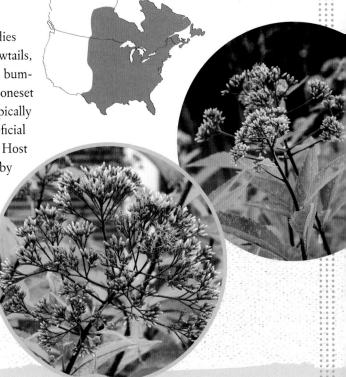

BLOOM TIME
Summer

FLOWER COLOR
Pink, purple

MAXIMUM HEIGHT
7 feet (2 m)

21. LOBELIA

(Lobelia spp.)

T HE NATIVE LOBELIAS feature showy flowers on a par with any introduced ornamentals for landscape appeal. The two commercially available species are among the larger members of the genus and produce the brightest blooms in partial shade. Both are adapted to damp soils, making them excellent for rain gardens and other temporarily wet situations. Honey bees can't "see" the flowers of cardinal flower: the blooms would appear black to them. When they do find the plant, however, they frantically mob it and rob it of nectar, because it is a prolific producer.

EXPOSURE
Sun to part shade

SOIL MOISTURE
Average to wet

RECOMMENDED SPECIES OR VARIETIES

Of the more than two dozen species in North America, only two, shown here, are regularly available: great blue lobelia (*Lobelia siphilitica*) and red-blossomed cardinal flower (*L. cardinalis*).

USES

Wildflower meadow/ prairie restoration

Wetland restoration

Ornamental

NOTABLE FLOWER VISITORS

Great blue lobelia is an exceptional bumble bee plant, attracting few other insects. Cardinal flower is pollinated by hummingbirds and is visited by a few butterflies, especially monarchs, for nectar. Honey bees' tongues are too short to extract nectar from the cardinal flower's blossom opening, so instead they insert their tongues into the base of the long blossoms, bypassing the reproductive structures of the flowers and robbing the nectar. Consequently, the flowers may be less attractive to the plant's legitimate pollinators, with unknown consequences for the plant's chances of reproducing.

BLOOM TIME	FLOWER COLOR	MAXIMUM HEIGHT
Summer	Red, blue	4 feet (1.2 m)

22. LUPINE
(Lupinus spp.)

WESTERN DESERTS, salty coastal dunes, and cool mountain stream banks all host uniquely adapted lupine species, sometimes in close proximity to one another. Several dozen different species are found in the United States and Canada, with the overwhelming majority located in the West, especially California. Lupines include both small annuals and large shrubby perennials. Most produce high-quality pollen, although they're not esteemed as nectar plants.

EXPOSURE
Sun to part shade

SOIL MOISTURE
Average to dry

RECOMMENDED SPECIES OR VARIETIES

In the eastern United States and Canada, perennial lupine (*Lupinus perennis*) is the most widely distributed species, best in deep, sandy soils. This species is the host plant for the endangered Karner blue butterfly (*Lycaeides melissa samuelis*), and the loss of lupine from the landscape has pushed the butterfly close to extinction. Silvery lupine (*L. argenteus*) is one of the most widely distributed species throughout the Great Basin, Rocky Mountains, and desert Southwest.

In California, yellow-flowered annual golden lupine (*L. densiflorus*) has performed well in Xerces Society pollinator meadows and cover crops at farms across the state, and the perennial summer lupine (*L. formosus*) has proven an excellent bumble bee plant. In rainy areas of the Pacific Northwest, riverbank lupine (*L. rivularis*) rapidly, almost aggressively, establishes itself in seeded pollinator meadows, successfully crowding out weeds. It mixes well with Puget Sound gumweed, and despite the common name, doesn't require riverbanks as habitat.

USES

Hedgerow

Wildflower meadow/ prairie restoration

Cover crop

Pollinator nesting material or caterpillar host plant

Ornamental

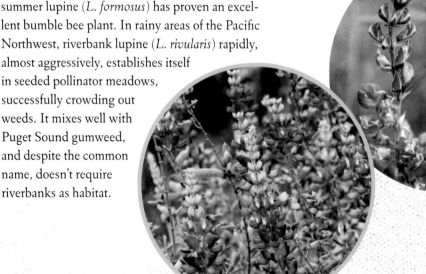

BLOOM TIME
Spring to summer

FLOWER COLOR
Blue, purple, white, yellow

MAXIMUM HEIGHT
6 feet (1.8 m)

NOTABLE FLOWER VISITORS

Attracts bumble bees, some mason bees, and occasionally pollen-gathering honey bees. Host plant for caterpillars of many butterflies, including clouded sulphurs (*Colias philodice*), orange sulphurs (*Colias eurytheme*), Persius duskywing (*Erynnis persius*), wild indigo duskywing (*Erynnis baptisiae*), frosted elfin (*Callophyrys irus*), gray hairstreak (*Strymon melinus*), sooty hairstreak (*Satyrium fuliginosum*), Acmon blue (*Plebejus acmon*), arrowhead blue (*Glaucopsyche piasus*), the Boisduval's blue (*Plebejus icarioides*), eastern tailed-blue (*Cupido comyntas*), melissa blue (*Lycaeides melissa*), silvery blue (*Glaucopsyche lygdamus*), and the endangered Karner blue (*Lycaeides melissa samuelis*).

23. MEADOWFOAM

(*Limnanthes* spp.)

WELL-KNOWN TO BEEKEEPERS, meadowfoam produces honey with a flavor frequently compared to marshmallows or vanilla. In the wild these annuals are primarily found in temporarily flooded or vernal pools in California and southern Oregon. Despite that limited range and habitat requirement, meadowfoam is surprisingly adaptable and grows well in many areas along the West Coast. It is now a staple or "workhorse plant" in most Xerces pollinator seed mixes in the maritime Pacific Northwest, reseeding very well and integrating successfully with species such as farewell to spring (*Clarkia amoena*) and Oregon phacelia (*Phacelia nemoralis*). In addition to seed mixes, meadowfoam is a good choice for a cover crop or as a temporary insectary planting sown between annual row crops.

EXPOSURE
Sun

SOIL MOISTURE
Average to wet

RECOMMENDED SPECIES OR VARIETIES

White meadowfoam (*Limnanthes alba*) and Douglas meadowfoam (*L. douglasii*) are available from some wildflower seed producers. The two resemble each other very closely except for flower color. White meadowfoam is also grown as a high-value oilseed crop; seed producers in the Northwest have attempted to limit commercial availability of the plant in order to maintain control over the supply.

NOTABLE FLOWER VISITORS

Attracts huge numbers of honey bees, blue orchard bees (*Osmia lignaria*), the Oregon berry bee (*O. aglaia*), mining bees (*Andrena* spp.), and many types of hoverflies.

USES

Wildflower meadow/ prairie restoration

Wetland restoration

Cover crop

Ornamental

BLOOM TIME	FLOWER COLOR	MAXIMUM HEIGHT
Spring	White, yellow	1 foot (30 cm)

24. MILKWEED
(*Asclepias* spp.)

CRITICALLY IMPORTANT for both the diversity and the abundance of pollinators they support, milkweeds are also among the most diverse and widely distributed native wildflowers in North America. They include nearly leafless cactuslike plants found in the desert Southwest and swamp-adapted species in the Florida Everglades, as well as the common milkweed growing along roadsides from New England to North Dakota. The honey of milkweeds is almost white with a very mild flavor and reported yields of 50 to 100 pounds per colony. The average sugar concentration in the nectar of some species has been reported at 37%.

EXPOSURE
Sun

SOIL MOISTURE
Wet to dry

RIGHT Brown-belted bumble
bee on milkweed

NOTABLE FLOWER VISITORS

Attracts many bees, wasps, flies, butterflies
such as swallowtails and fritillaries, and even
hummingbirds. An interesting characteristic of
milkweed flowers are their **pollinia**, or pollen-
bearing sacs. These hook themselves onto the
legs and tongues of flower-visiting insects and
are usually released only when the insect visits
another milkweed flower and inserts the pollinia
into a slot, much like placing a key in a lock.
Caterpillar host plant for monarch (*Danaus plex-
ippus*), queen (*D. gilippus*), and soldier butterflies
(*D. eresimus*) as well as the unexpected cycnia
(*Cycnia inopinatus*), dogbane tiger (*C. tenera*), and
milkweed tussock (*Euchaetes egle*) moths.

USES

Hedgerow

**Wildflower meadow/
prairie restoration**

Wetland restoration

Farm buffer/filter strip

**Pollinator nesting material
or caterpillar host plant**

Ornamental

RECOMMENDED SPECIES OR VARIETIES

Butterfly weed (*Asclepias tuberosa*), swamp milkweed (*A. incarnata*); many other locally adapted
species.

BLOOM TIME	FLOWER COLOR	MAXIMUM HEIGHT
Summer	White, purple, pink, orange, green	5 feet (1.5 m)

25. MOUNTAINMINT
(*Pycnanthemum* spp.)

THESE NATIVE MINTS, close relatives of beebalm, have a strong and very pleasant odor when the foliage is crushed. Many anecdotal reports and long-term observation by Xerces staff assert that these plants are a honey bee favorite. One Xerces staff member has recorded extremely good honey yields near a field of mountainmint grown for commercial seed production.

EXPOSURE
Sun

SOIL MOISTURE
Average to dry

RECOMMENDED SPECIES OR VARIETIES

Virginia mountainmint (*Pycnanthemum virginianum*) is the most widely available species, but roughly a dozen other locally adapted species can be found in eastern North America. Short-toothed or clustered mountainmint (*P. muticum*) is the most ornamental of the bunch, popular for perennial plantings and attractive to butterflies. Uncommon, but worth seeking out, is the California native Sierra mint (*P. californicum*).

NOTABLE FLOWER VISITORS

The shallow nectaries of mountainmint and its seemingly abundant nectar attract an amazing parade of bees, beneficial solitary wasps, flies, beetles, and small butterflies such as hairstreaks.

USES

**Wildflower meadow/
prairie restoration**

Ornamental

Edible/herbal/medicinal

BLOOM TIME	FLOWER COLOR	MAXIMUM HEIGHT
Late summer	White	4 feet (1.2 m)

26. NATIVE THISTLE
(*Cirsium* spp.)

THESE ATTRACTIVE PLANTS should not be confused with invasive alien species such as Canada thistle and bull thistle. Several dozen wild, nonweedy thistles once existed as part of most prairie, meadow, grassland, and desert ecosystems in North America. Some are short-lived plants that depend on soil disturbance to germinate and grow, only to be crowded out by longer-lived, more aggressive species.

Native thistles support pollinators and songbirds such as indigo buntings. Unfortunately, thistles are disappearing due to habitat loss, eradication efforts targeting invasive alien thistles, and exotic thistle-feeding insects released to control Canada thistle. Thistle honey is clear or white and has been compared to basswood honey in flavor.

EXPOSURE
Sun to part shade

SOIL MOISTURE
Average to wet

RECOMMENDED SPECIES OR VARIETIES

In eastern North America, field thistle (*Cirsium discolor*) is probably the most adaptable and widespread species. Tall thistle (*C. altissimum*), typically a woodland-edge species, can reach amazing heights of 8 feet or more in optimal soils. A western species, cobweb thistle (*C. occidentale*), is covered with white hairs, tinting the entire plant a bright white that contrasts strikingly with its blood-red flowers. Unfortunately, seed of these and most native thistles is available only from a few specialty nurseries.

NOTABLE FLOWER VISITORS

Attracts bumble bees, including the very large black-and-gold bumble bee (*Bombus auricomus*), leafcutter bees (*Megachile* spp.), and butterflies such as monarchs and swallowtails; occasional hummingbirds. A food source for caterpillars of painted lady (*Vanessa cardui*), swamp metalmark (*Calephelis muticum*), mylitta crescent (*Phyciodes mylitta*), and California crescent (*Phyciodes orseis*) butterflies. Additional specialist bees include *Melissodes desponsa*, *Osmia chalybea*, and *O. texana*.

USES

Wildflower meadow/ prairie restoration

Pollinator nesting material or caterpillar host plant

Ornamental

Edible/herbal/medicinal

RIGHT American bumble bee foraging native thistle

BLOOM TIME
Summer to fall

FLOWER COLOR
Pink, white, red

MAXIMUM HEIGHT
8 feet (2.4 m)

27. PENSTEMON
(*Penstemon* spp.)

SOMETIMES CALLED BEARDTONGUES, many penstemon species occur in most parts of North America, and nearly all are excellent pollinator plants, with attractive flowers. The types of pollinators they attract vary depending on the species; some showy red-flowered penstemons attract hummingbirds, and others support sphinx moths. The largest are tall enough to plant on the edges of hedgerows in the West. Smaller species work well as meadow plants, especially among smaller grasses that will not shade them out. The average sugar concentration in the nectar of some penstemon species has been reported at 37%.

EXPOSURE
Sun to part shade

SOIL MOISTURE
Average to dry

RECOMMENDED SPECIES OR VARIETIES

In the West, Venus penstemon (*Penstemon venustus*), Palmer's penstemon (*P. palmeri*), Eaton's penstemon (*P. eatonii*); in the East, smooth penstemon (*P. digitalis*), large-flowered penstemon (*P. grandiflorus*).

NOTABLE FLOWER VISITORS

Attracts honey bees, vast numbers of native bees. Both *Osmia distincta* and *Pseudomasaris occidentalis*, an unusual pollen-collecting wasp, prefer penstemons. Various penstemons are host plants for caterpillars of arachne (*Poladryas arachne*) and variable checkerspot (*Euphydryas chalcedona*) butterflies.

USES

Hedgerow

Wildflower meadow/ prairie restoration

Pollinator nesting material or caterpillar host plant

Ornamental

BLOOM TIME
Summer

FLOWER COLOR
White, pink, purple, red

MAXIMUM HEIGHT
4 feet (1.2 m)

28. PHACELIA
(*Phacelia* spp.)

THE ODD-LOOKING FIDDLENECK FLOWERS of phacelia, and the itchy hairs that often cover the leaves and stems, do not endear these plants to many people. Their abundant nectar, however, makes them a favorite of bees. One species, lacy phacelia (*Phacelia tanacetifolia*), is sometimes mass-planted for honey bee forage and commonly planted as a cover crop in Europe, where it is an introduced species. Note that lacy phacelia also attracts lygus bug — a serious pest — and so should not be planted near susceptible crops such as strawberries. Although various *Phacelia* species are found across North America, the West is where they are found in greatest abundance.

EXPOSURE
Sun

SOIL MOISTURE
Average to dry

RECOMMENDED SPECIES OR VARIETIES

Annual lacy phacelia, sometimes known as scorpionweed, is the best known. The desert annual California bluebells (*P. campanularia*) is available as low-cost bulk wildflower seed and can be grown in many climates, although it favors dry, sandy soils. In the Pacific Northwest, perennial Oregon phacelia (*P. nemoralis*) tolerates partial shade and attracts remarkable bumble bee activity, especially where clusters of the plant grow in semiopen meadows.

NOTABLE FLOWER VISITORS

Attracts common species, mostly bumble bees, honey bees, mason bees, and syrphid flies. The bee activity, especially on lacy phacelia, tends to amaze observers seeing it for the first time.

USES

Wildflower meadow/ prairie restoration

Cover crop

Ornamental

BLOOM TIME
Spring

FLOWER COLOR
Purple, white, blue

MAXIMUM HEIGHT
2 feet (0.6 m)

29. PRAIRIE CLOVER
(*Dalea* spp.)

PRAIRIE CLOVERS are relatively slow growing and eagerly devoured by livestock — traits that, along with loss of prairie habitat, have made them relatively scarce across much of their historic range. Where these long-blooming plants exist in good numbers, they are worthy of special interest for beekeepers. In large-scale seed production fields at native plant nurseries, Xerces Society members have witnessed huge numbers of honey bees on these plants, seemingly ignoring any other plants in the area, and honey yields estimated to exceed 100 pounds per hive. In addition to their honey bee value, the flowers of some species, such as purple prairie clover, deliver an incredibly bright, cheerful ornamental show of almost fluorescent pinkish purple blossoms atop delicate fine-leaved stems.

EXPOSURE
Sun

SOIL MOISTURE
Average to dry

RECOMMENDED SPECIES OR VARIETIES

Purple prairie clover (*Dalea purpurea*) is a widespread perennial across much of North America, with prolific purple flowers. It should be considered a "must-have" in pollinator conservation efforts where it is native and appropriate to the site. White prairie clover (*D. candida*) is less showy than purple prairie clover but can tolerate drier sites. Many other species can be found across the West, especially in Texas and the Southwest.

NOTABLE FLOWER VISITORS

Attracts honey bees and bumble bees (including the endangered rusty patched bumble bee, *Bombus affinis*) in large numbers. Specialist pollinators include the polyester bees *Colletes albescens*, *C. susannae*, *C. wilmattae*, and *C. robertsonii*. Host plant for caterpillars of southern dogface (*Zerene cesonia*), clouded sulphur (*Phoebis sennae*), marine blue (*Leptotes marina*), and Reakirt's blue (*Hemiargus isola*) butterflies.

USES

Wildflower meadow/
prairie restoration

Rangeland/pasture

Farm buffer/filter strip

Pollinator nesting material
or caterpillar host plant

Ornamental

BLOOM TIME	FLOWER COLOR	MAXIMUM HEIGHT
Summer	White, purple	3 feet (0.9 m)

30. PURPLE CONEFLOWER

(*Echinacea* spp.)

WITH COLORFUL DAISYLIKE FLOWERS, purple coneflowers make beautiful additions to ornamental gardens as well as wildflower meadows and butterfly gardens. All purple coneflowers attract a variety of bees and butterflies. Honey bees are common visitors; when foraging on purple coneflowers they are frequently observed with full pollen baskets, suggesting that the plants are important for supplementing and diversifying bee diets.

EXPOSURE
Sun

SOIL MOISTURE
Average

**Wildflower meadow/
prairie restoration**

Rangeland/pasture

**Pollinator nesting material
or caterpillar host plant**

Ornamental

Edible/herbal/medicinal

RECOMMENDED SPECIES OR VARIETIES

Common purple coneflower (*E. purpurea*) tends to be the most adaptable and commercially available of the various coneflowers. Pale purple coneflower (*Echinacea pallida*) and narrow-leaved coneflower (*E. angustifolia*) also attract many pollinators. All are relatively slow growing and take several years to begin flowering when grown from seed, but they can be extremely long-lived plants under optimal conditions.

NOTABLE FLOWER VISITORS

Attracts bumble bees, sweat bees, and various so-called "sunflower bees" in the genera *Diadasia*, *Melissodes*, and *Svastra*. Specialists include the sunflower leafcutter bee (*Megachile pugnata*) and a mining bee (*Andrena helianthiformis*). Many butterflies also visit for nectar, including monarchs, swallowtails, and sulphurs. Host plant for caterpillars of silvery checkerspot butterfly (*Chlosyne nycteis*).

RIGHT Mining bees on purple coneflower

BLOOM TIME	FLOWER COLOR	MAXIMUM HEIGHT
Summer	Purple	4 feet (1.2 m)

31.

RATTLESNAKE MASTER, ERYNGO

(*Eryngium* spp.)

GLOBELIKE BLOSSOMS and tough, sometimes spiny foliage give these members of the carrot family a striking appearance. The genus includes both perennial and annual species that are adaptable to a variety of habitats. The honey of these plants is described as dark in color and pleasantly flavored.

EXPOSURE
Sun to part shade

SOIL MOISTURE
Wet to dry

RECOMMENDED SPECIES OR VARIETIES

Perennial rattlesnake master (*Eryngium yuccifolium*) is a prairie native commonly available from native plant nurseries. Its scientific name refers to the plant's yuccalike leaves, while the common name supposedly refers to its historic use as a home remedy for rattlesnake bites (not recommended!). In the southern plains, Leavenworth's eryngo (*E. leavenworthii*) is a beautiful annual with lavender flowers that rival any cultivated ornamental. Across much of the United States, the introduced sea holly (*E. maritimum*) is a common flower garden plant with blue foliage and blossoms; like its native relatives it attracts many bees.

NOTABLE FLOWER VISITORS

Attracts many small sweat bees, syrphid flies, beneficial wasps, and attracts black-and-gold bumble bees (*Bombus auricomus*). Caterpillar host plant for the endangered rattlesnake borer moth (*Coleotechnites eryngiella*). The hollow stems of rattlesnake master are slow to break down and provide nest sites for various wood-nesting bees.

USES

Wildflower meadow/ prairie restoration

Pollinator nesting material or caterpillar host plant

Ornamental

RIGHT Southern plains bumble bee on rattlesnake master

BLOOM TIME	FLOWER COLOR	MAXIMUM HEIGHT
Summer	White, blue, purple	6 feet (1.8 m)

32.
ROCKY MOUNTAIN BEE PLANT

(*Cleome* spp.)

L ANKY BUT DAZZLING, this purple-flowered western dryland plant has long been considered an important honey plant, with reports of honey yields greater than 100 pounds per colony over a 10-day period, and 2 to 3 supers (surplus honey boxes) added per colony over a 3-week period. The resulting greenish white honey has been described as variable, with an unpleasant flavor that improves with age. Since the plant and its close relatives can tolerate soil and climate conditions where few other plants grow, they are regionally important. The average sugar concentration in the nectar of Rocky Mountain bee plant has been reported at 21 to 29%.

EXPOSURE
Sun

SOIL MOISTURE
Wet to dry

RECOMMENDED SPECIES OR VARIETIES

Annual pink-flowered Rocky Mountain bee plant (*Cleome serrulata*) grows in disturbed semimoist rangeland soils. Its close relative, the annual yellow bee plant (*C. lutea*), grows in even harsher conditions, including areas of alkali soils with less than 10 inches of annual rainfall. In the deserts of southern California, the shrub bladderpod (*Peritoma arborea*, formerly *Isomelis arborea*) is a close relative that also survives on minimal annual rainfall, while blooming abundantly and almost continuously throughout the year. It is very attractive to honey bees and an excellent choice for farm hedgerows in drought-prone areas.

NOTABLE FLOWER VISITORS

Best known for attracting honey bees, but monarch butterflies and hummingbirds also visit the flowers for nectar. Scientists speculate that yellow bee plant was a principal food source for the alkali bee (*Nomia melanderi*, an important crop pollinator) before the introduction of alfalfa.

USES

Hedgerow

Wildflower meadow prairie restoration

Rangeland/pasture

Ornamental

BLOOM TIME	FLOWER COLOR	MAXIMUM HEIGHT
Summer	Purple, pink, yellow	4 feet (1.2 m)

33. SALVIA
(*Salvia* spp.)

A DIVERSE GROUP OF ANNUALS and perennials, natives and nonnatives, many salvias have been cultivated as ornamental garden plants. Several are considered important honey bee forage plants while others, especially red-flowered species, are more likely to attract hummingbirds. Though salvias are commonly called sage, they should not be confused with sagebrush (*Artemisia* spp.), a group of plants that attract few bees. The honey is pale yellow and slow to granulate. The average sugar concentration in the nectar of black sage (*Salvia mellifera*) has been reported at 44%.

EXPOSURE
Sun

SOIL MOISTURE
Average to dry

RECOMMENDED SPECIES OR VARIETIES

Many salvias succeed in pollinator habitat projects. Though not native, the garden herb sage (*Salvia officinalis*) is adapted to moderate coastal climates and attracts many bees. The California native black sage (*S. mellifera*) is considered a premium nectar plant, producing top-quality, heavy white honey that almost never granulates. Its large, shrubby size makes it a convenient hedgerow plant on California farms. The annual chia (*S. hispanica*) — yes, the "*cha-cha-cha*" Chia Pet plant — is another good honey bee forage plant. It requires relatively even day lengths to promote flowering, so its value is probably restricted to the extreme Southwest borderlands. Its close relative desert chia (*S. columbariae*) grows farther north into Nevada and California. In the southern plains, the perennial blue sage (*S. azurea*) is a beautiful component of remnant prairies and roadsides.

NOTABLE FLOWER VISITORS

Attracts honey bees, many other insects, and hummingbirds. Some species are host plants for caterpillars of the elegant sphinx moth (*Sphinx perelegans*).

USES

Hedgerow

Wildflower meadow/ prairie restoration

Rangeland/pasture

Cover crop

Ornamental

Edible/herbal/medicinal

BLOOM TIME
Summer

FLOWER COLOR
Blue, purple, pink, white, red

MAXIMUM HEIGHT
4 feet (1.2 m)

34.

SELFHEAL

(Prunella vulgaris)

SELFHEAL IS BROADLY DISTRIBUTED across the Northern Hemisphere. Though native to North America, many of the individual plants observed, especially in eastern North America, are weeds of European origin, with reduced flowers. The plant competes well against other vegetation, especially weedy grasses that choke out other, more fragile wildflowers. When seeded at high enough densities, selfheal will form robust colonies of plants that tolerate occasional mowing and vehicle traffic. These characteristics make it a suitable forage plant for honey bees and other beneficial insects in orchard and vineyard understories.

EXPOSURE
Sun to part shade

SOIL MOISTURE
Average to wet

RECOMMENDED SPECIES OR VARIETIES

The native North American selfheal is taxonomically defined as the subspecies *lanceolata*. Its introduced counterpart (subspecies *vulgaris*) has typically smaller flowers and a low, creeping growth habit. Seed of the native North American subspecies is typically available only in the Pacific Northwest.

ABOVE Bumble bee on selfheal

NOTABLE FLOWER VISITORS

Attracts bumble bees most commonly, also honey bees. Many other bee species probably visit as well.

USES

Wildflower meadow/ prairie restoration

Rangeland/pasture

Farm buffer/filter strip

Cover crop

Edible/herbal/medicinal

BLOOM TIME
Early summer

FLOWER COLOR
Blue

MAXIMUM HEIGHT
1 foot (30 cm)

35. SNEEZEWEED

(*Helenium* spp.)

WIDELY DISTRIBUTED and commonly available, sneezeweed (*Helenium autumnale*) prefers wet, disturbed sites such as sandy riverbanks and backwater shorelines. This adaptable plant is a good choice for swampy, seasonally flooded areas of a farm or a roadside. Honey produced from sneezeweed is reportedly so bitter that it ruins any other honey mixed with it. While sneezeweed is a very good honey producer in some areas of the South, beekeepers typically leave its honey on hives as a source of winter food for the bees.

EXPOSURE
Sun

SOIL MOISTURE
Average to wet

RECOMMENDED SPECIES OR VARIETIES

While various ornamental cultivars of sneezeweed have been developed for gardens, the wild species probably attracts the greatest number of pollinators. A close relative found in parts of Texas and Oklahoma, yellowdicks (*Helenium tenuifolium*) is well known as a plant that attracts honey bees in abundance but produces bitter honey.

NOTABLE FLOWER VISITORS

Attracts many bees, including honey bees, bumble bees, and leafcutter bees.

USES

Wildflower meadow/prairie restoration

Wetland restoration

Farm buffer/filter strip

Ornamental

BLOOM TIME	FLOWER COLOR	MAXIMUM HEIGHT
Summer to autumn	Yellow	5 feet (1.6 m)

36. SPIDERWORT
(*Tradescantia* spp.)

TYPICALLY, THE BLOOMS OF SPIDERWORT open in the early morning but begin to wilt by midday, leaving behind a wet blue remnant that can stain clothing. Spiderwort flowers reportedly do not produce nectar, although the pollen is apparently valuable enough to attract significant numbers of bees and pollen-feeding flies. In observations by Xerces Society members, spiderwort's attractiveness to bees increases with the number of individual plants, so for best effect it should probably be mass-planted.

EXPOSURE
Sun to part shade

SOIL MOISTURE
Average

RECOMMENDED SPECIES OR VARIETIES

Several species are available from native plant nurseries and seed dealers, including Ohio spiderwort (*Tradescantia ohiensis*), prairie spiderwort (*T. occidentalis*), and Virginia spiderwort (*T. virginiana*). All are similar in appearance. Ornamental cultivars tend to attract few bees, in our experience.

NOTABLE FLOWER VISITORS

Attracts pollen-collecting bumble bees and honey bees as well as pollen-feeding flies such as syrphids.

USES

Wildflower meadow/ prairie restoration

Ornamental

BLOOM TIME
Late spring to early summer

FLOWER COLOR
Blue, purple

MAXIMUM HEIGHT
2 feet (0.6 m)

37. SUNFLOWER

(*Helianthus* spp.)

PERHAPS NO FLOWER REPRESENTS SUMMER in North America as well as the cheerful, ubiquitous, and multi-formed sunflower, one of our most diverse and widespread natives. Many sunflower species and horticultural varieties are available, and all attract a tremendous diversity of insects, including bees, wasps, flies, butterflies, and pollen-feeding soldier beetles. When planting the common annual sunflower (*Helianthus annuus*), avoid "pollenless" or double-petaled ornamental varieties. The average sugar concentration in the nectar of sunflower species reportedly ranges from 31 to 49%.

EXPOSURE
Sun to shade

SOIL MOISTURE
Average to dry

RECOMMENDED SPECIES AND VARIETIES

Woodland sunflower (*Helianthus divaricatus*) is suited to shady locations in the eastern United States and Canada, while the smaller prairie sunflower (*H. petiolaris*) prefers full sun and sandy soils. Tall Maximilian sunflower (*H. maximiliani*) is suited to dry sunny sites, especially in the Great Plains, where it can be very long lived once established. Other species occur in the Great Basin and California, all of which are excellent pollinator plants. Taller species in the West have established well in hedgerows planted by the Xerces Society. The common annual sunflower (*H. annuus*) is a valuable addition to warm-season cover-crop seed mixes.

USES

Hedgerow

Wildflower meadow/ prairie restoration

Reclaimed industrial land/ tough sites

Rangeland/pasture

Cover crop

Pollinator nesting material or caterpillar host plant

Ornamental

Edible/herbal/medicinal

NOTABLE FLOWER VISITORS

All bees in the genera *Diadasia, Melissodes, Eucera,* and *Svastra* are common specialists. Other specialists include *Andrena accepta, A. aliciae, A. helianthi, Dufourea marginatus, Melissodes agilis, Pseudopanurgus rugosus, Perdita bequaerti, Paranthidium jugatorium, Dieunomia heteropoda* (the largest member of the Halictidae bee family in the United States), and the sunflower leafcutter bee (*Megachile pugnata*). Leafcutters may use the dead hollow stems of large sunflowers for nesting. Host plant for caterpillars of silvery checkerspot (*Chlosyne nycteis*) and the bordered patch (*C. lacinia*) butterflies.

BLOOM TIME	FLOWER COLOR	MAXIMUM HEIGHT
Late summer to autumn	Yellow, orange	8 feet (2.4 m)

38. WATERLEAF

(Hydrophyllum spp.)

IN MOST WATERLEAF SPECIES, the flower clusters rise from long stalks, and the deeply cut leaves are dappled by lighter-colored marks as though stained by water. These plants strongly favor shaded locations and moist soils, making them useful for shade gardens, in forested understory meadows, and as lower canopy plants within hedgerows.

EXPOSURE	SOIL MOISTURE
Shade	Average

RECOMMENDED SPECIES OR VARIETIES

The common eastern waterleaf, also known as Virginia waterleaf (*Hydrophyllum virginianum*), is the most widespread, commonly available, and probably the most attractive to bees. It is rarely available as seed, but plants can be purchased from some native plant nurseries. Under favorable conditions it will spread.

NOTABLE FLOWER VISITORS

Attracts bumble bees (especially newly emerged spring queens) and the blue orchard bee (*Osmia lignaria*). *Andrena geranii* is a specialist of eastern waterleaf.

USES

Hedgerow

Reforestation/shade garden

Ornamental

BLOOM TIME	FLOWER COLOR	MAXIMUM HEIGHT
Spring	Purple, pink, white	1 foot (30 m)

39.
WILD BUCKWHEAT
(*Eriogonum* spp.)

IN CALIFORNIA, the Great Basin, and the Southwest, wild buckwheat is a "must-have" plant for pollinator gardens. Few other dryland plant groups flower as reliably and prolifically as wild buckwheat in summer and fall, or attract the sheer numbers and diversity of small wild bees and butterflies. In particular, a seemingly endless parade of various blue butterflies (Lycaenids) use wild buckwheats as a food source during both their adult and larval stages. Sugar concentration in the nectar of some wild buckwheat species has been reported as high as 58%.

EXPOSURE
Sun

SOIL MOISTURE
Dry

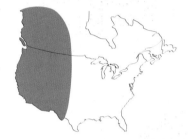

RECOMMENDED SPECIES OR VARIETIES

*E*riogonum umbellatum is highly versatile and wide-ranging. *E. wrightii* is fantastic for xeric conditions and rock gardens from southern California through west Texas. *E. ovalifolium* is highly adaptable across the western mountain regions, ranging from southern Canada to Arizona.

NOTABLE FLOWER VISITORS

Extremely important for several butterfly caterpillars; the flowers are an important floral source for those same butterflies. Some caterpillars known to feed on wild buckwheat include the acmon blue (*Plebejus acmon*), blue copper (*Lycaena heteronea*), gorgon copper (*Lycaena gorgon*), bramble hairstreak (*Callophrys dumetorum*), desert green hairstreak (*Callophrys sheridanii*), lupine blue (*Plebejus lupini*), Mormon metalmark (*Apodemia mormo*), Rocky Mountain dotted blue (*Euphilotes ancilla*), Sonoran metalmark (*Apodemia mejicanus*), western green hairstreak butterflies (*Callophrys affinis*), and the electra buckmoth (*Hemieuca electra*).

USES

Hedgerow

Wildflower meadow/ prairie restoration

Rangeland/pasture

Pollinator nesting material or caterpillar host plant

Ornamental

BLOOM TIME
Summer to autumn

FLOWER COLOR
White, yellow, pink

MAXIMUM HEIGHT
6 feet (1.8 m)

40. WILD GERANIUM

(*Geranium* spp.)

VARIOUS BEES VISIT most of the perennial and annual species of wild geraniums. In general, these plants are smaller in stature and less showy than ornamental geraniums, and many thrive in shade or forest understory locations.

EXPOSURE
Shade to part shade

SOIL MOISTURE
Average

RECOMMENDED SPECIES OR VARIETIES

In the West, sticky purple geranium (*Geranium viscosissimum*); in the East, Bicknell's cranesbill (*G. bicknellii*) and spotted geranium (*G. maculatum*).

NOTABLE FLOWER VISITORS

Attracts mason bees (*Osmia* spp.), mining bees (*Andrena* spp.), wood-nesting green metallic sweat bees (*Augochlora* spp.), and bumble bees (*Bombus* spp.). One mining bee, *Andrena distans*, is a known **oligolectic** (pollen-gathering specialist) of wild geranium. In the West, the green metallic mason bee (*Osmia aglaia*) has been observed collecting leaf pieces from various wild geraniums to build nest cells.

USES

Hedgerow

Reforestation/shade garden

Pollinator nesting material or caterpillar host plant

Ornamental

BLOOM TIME	FLOWER COLOR	MAXIMUM HEIGHT
Spring	White, yellow, pink	2 feet (0.6 m)

41. WILD INDIGO

(*Baptisia* spp.)

TALL, FLESHY PLANTS of the central and eastern prairie regions, with vivid, lupinelike flower spikes, most wild indigos are slow growing, sometimes taking several years to appear in restored prairies after they are planted from seed. The root systems tend to be deep and extensive, and once established most species can be extremely long lived.

EXPOSURE
Sun

SOIL MOISTURE
Average

RECOMMENDED SPECIES OR VARIETIES

Blue wild indigo (*Baptisia australis*), white wild indigo (*B. alba*). The faster-growing horseflyweed (*B. tinctoria*) has smaller yellow flowers that are more attractive to honey bees. It was once known to rapidly recolonize burned areas in the eastern United States, where it was a valuable honey plant.

USES

Wildflower meadow/
prairie restoration

Pollinator nesting material
or caterpillar host plant

Ornamental

NOTABLE FLOWER VISITORS

Long-tongued bees most easily access nectar and pollen from the deep flowers, especially queen bumble bees. Host plant for caterpillars of clouded sulphur (*Colias philodice*) and wild indigo duskywing (*Eynnis baptisiae*) butterflies, as well as black-spotted prominent (*Dasylophus anguina*) and genista broom (*Uresiphita reversalis*) moths. Only horseflyweed is a honey bee forage plant, with yields of over 150 pounds of honey per hive reported in New York during the 1920s.

RIGHT Dusted skipper (top)
and southern cloudywing
butterflies on wild indigo

BLOOM TIME
Spring to summer

FLOWER COLOR
Blue, white, yellow, rose

MAXIMUM HEIGHT
5 feet (1.5 m)

42.

WINGSTEM
(*Verbesina* spp.)

ALTHOUGH TALL and sometimes described as "gangly," wingstem's interesting stem shape and bright flowers easily lend themselves to use in ornamental shade gardens. Under the name "golden honey plant," common wingstem seed was once produced and marketed specifically to beekeepers in the central United States, with seed advertisements appearing in beekeeping magazines even in the early 1900s. The honey is described as being exceptional in quality, gold in color, and can be harvested in enormous quantities when good growing conditions exist. One report from Kentucky in the 1920s described more than 2,000 pounds of wingstem honey being harvested from 22 hives. Most species grow best in damp, slightly shaded sites such as drainage ditches, stream banks, floodplains, and riparian buffers.

EXPOSURE
Shade to part sun

SOIL MOISTURE
Wet to average

RECOMMENDED SPECIES OR VARIETIES

Various species are widely distributed, and the most common tend also to be the best nectar and pollen plants. Commercial seed sources are limited but slowly increasing in availability. Look for common wingstem (*Verbesina alternifolia*), golden crownbeard (*V. encelioides*), and white crownbeard (*V. virginica*).

NOTABLE FLOWER VISITORS

Although sometimes considered weedy, various species of wingstem (also called crownbeard and frostweed, depending on the species) attract extremely large numbers of beneficial insects. Its hollow stems provide suitable nesting habitat for leafcutter and mason bees, and the plant is a caterpillar host for the bordered patch butterfly and gold moth.

ABOVE Honey bee on wingstem

USES

Reforestation/shade garden

Wetland restoration

Farm buffer/ filter strip

Pollinator nesting material or caterpillar host plant

Ornamental

BLOOM TIME
Summer

FLOWER COLOR
Yellow, white

MAXIMUM HEIGHT
8 feet (2.4 m)

43. WOOD MINT

(Blephilia ciliata)

THE GORGEOUS STACKED PURPLE FLOWERS of wood mint are sometimes described as "pagoda-like." Blooming prolifically for many weeks in midsummer, this plant deserves more attention by pollinator gardeners, both as a prolific nectar plant and as a beautiful ornamental. It also has the notable ability to thrive in dry shady locations where many other plants would fail. Wood mint spreads easily by rhizomes and expands slowly into clumps wherever space is available. It mixes well with **spotted wild geranium** (see page 106) and **Virginia waterleaf** (see page 102) to provide a diverse and attractive pollinator garden for shady yards or forest clearings.

EXPOSURE
Part shade to shade

SOIL MOISTURE
Dry

RECOMMENDED SPECIES OR VARIETIES

No ornamental cultivars of wood mint exist.

NOTABLE FLOWER VISITORS

Visited extensively by bumble bees, honey bees, and leafcutter bees.

USES

Reforestation/shade garden

Ornamental

Edible/herbal/medicinal

LEFT Bumble bee on wood mint

BLOOM TIME
Late spring and summer

FLOWER COLOR
White, blue, purple

MAXIMUM HEIGHT
2 feet (0.6 m)

Native Trees
and Shrubs

Native woody plants provide important resources for wild and managed pollinators. Moreover, tree and shrub plantings may be designed for multiple additional purposes, such as the health of wildlife, soil, and water. Here are just some of the shrub and tree species that you might want to consider, staying mindful of "right plant, right place."

44. ACACIA

(*Senegalia* spp.)

THESE TOUGH, SPINY, DROUGHT-ADAPTED TREES yield honey in great abundance and are famous in Texas for their benefits to pollinators. Acacias also support birds by offering seeds for foraging and sites for nesting, and in hot, sunny environments they are an excellent summer shade plant for humans and livestock alike.

EXPOSURE

Part shade to sun

SOIL MOISTURE

Dry

RECOMMENDED SPECIES OR VARIETIES

Related acacias from Australia and elsewhere are found in the southwestern states, but they attract few bees. Look instead for the native huajillo (*Senegalia berlandieri*) and catclaw (*S. greggii*), two trees with well-deserved reputations for attractiveness to bees and butterflies.

NOTABLE FLOWER VISITORS

Attracts honey bees and various native bees. Native acacia trees are thought to be important caterpillar host plants for butterflies such as the marine blue (*Leptotes marina*), Reakirt's blue (*Echinargus isola*), silver-spotted skipper (*Epargyreus clarus*), outis skipper (*Cogia outis*), and Mexican yellow (*Eurema mexicana*).

USES

Hedgerow

Rangeland/
pasture

Pollinator nesting
material or caterpillar
host plant

Ornamental

BLOOM TIME	FLOWER COLOR	MAXIMUM HEIGHT
Spring, summer, fall	Orange, yellow, white	36 feet (11 m)

45. BASSWOOD
(Tilia americana)

A COMMON SHADE TREE IN CITIES, basswood is an important nectar source for bees and a most useful plant for beekeepers. It produces a pale but richly flavored honey, sometimes said to hint of peppermint. Its short bloom period typically lasts 10 days or less, and the nectar flow is irregular, more abundant in some years than others. Production of flowers and nectar increases with soil fertility and sunlight, so new plantings should have rich soil and full sun.

The average sugar concentration in the nectar of basswood flowers has been reported at 33%, and colony honey yields as high as 200 pounds have been reported under favorable conditions. Interestingly, honey yields are larger from basswood growing in mixed populations as opposed to single-species plantations.

EXPOSURE
Sun to shade

SOIL MOISTURE
Average

RECOMMENDED SPECIES OR VARIETIES

Basswood's European cousin, littleleaf linden (*Tilia cordata*), is a popular shade tree and also valued as a nectar-producing tree among beekeepers.

USES

Reforestation/shade garden

Ornamental

Edible/herbal/medicinal

NOTABLE FLOWER VISITORS

A showstopper for attracting honey and bumble bees, metallic green sweat bees, flies, and many wasps. Although plants in this genus have been implicated in bee poisonings (due to toxic nectar), much debate exists on this subject, and it seems that the triggers for toxic nectar production are not well understood (although excessive dryness and drought may be a factor). Currently it is believed that European species can be more toxic to bees (especially bumble bees) than the native common basswood, although such poisoning events are extremely rare overall.

Along with nectar, basswoods attract large numbers of aphids, whose corresponding honeydew (sugary excrement) is eagerly collected by many bees and wasps. Caterpillar host plant for the mourning cloak (*Nymphalis antiopa*), red-spotted purple (*Limenitis arthemis astyanax*), and eastern tiger swallowtail (*Papilio glaucus*) butterflies.

BLOOM TIME
Early summer

FLOWER COLOR
White

MAXIMUM HEIGHT
80 feet (24 m)

46. BLACKBERRY, RASPBERRY

(Rubus spp.)

MANY NATIVE AND NONNATIVE bramble species and cultivated varieties are found across the United States and much of Canada. They are valued as forage plants for honey bees as well as for their usually tasty berries: beekeepers report more than 25 pounds of surplus honey per hive, especially in the Northwest, where forest clearings and roadsides support vast thickets of the invasive Himalayan blackberry (*Rubus armeniacus*). The honey from these plants is white or light amber, typically very thick, and slow to granulate. These plants bloom right after tree fruits but before white clover, making them an important resource in farmlands and anywhere honey bees are managed.

EXPOSURE
Shade to sun

SOIL MOISTURE
Dry to wet

RECOMMENDED SPECIES OR VARIETIES

The common red raspberries (*Rubus ideaeus* and *R. strigosus*) grow just about everywhere and are reliable nectar producers, especially on warm days preceded by cool nights. The authors have a personal bias toward the related black raspberries (*R. occidentalis*), which are also widely adapted, packed with antioxidants, and more complexly flavored — oh, and they attract bees too. Hybrids include the wonderfully delicious and prolific marionberry (*Rubus* 'Marion', a cross of 'Chehalem' and 'Olallie'), common in the Northwest, supporting bees and pie bakers alike.

Beyond these semidomestic options, it's worth noting that there are various wonderful truly wild members of the genus that are both beautiful ornamentals in their own right and are pollinator magnets. One such example is salmonberry (*R. spectabilis*), a West Coast native of cool forest stream banks. Featuring bright magenta flowers, it's a great option for cool, shady native plant gardens.

NOTABLE FLOWER VISITORS

Attracts honey bees, bumble bees (*Bombus* spp.), and mining bees (*Andrena* spp.). An excellent and reliable resource for mass-rearing mason bees (*Osmia* spp.). Small carpenter bees (*Ceratina* spp.) frequently visit *Rubus* flowers and favor the hollow or pithy stems as nest sites. Host plant for caterpillars of echo azure butterfly (*Celastrina echo*) and Io moth (*Automeris io*).

USES

Hedgerow

Reforestation/shade garden

Pollinator nesting material or caterpillar host plant

Ornamental

Edible/herbal/medicinal

BLOOM TIME
Spring to summer

FLOWER COLOR
White, pink

MAXIMUM HEIGHT
12 feet (3.7 m)

47.

BLACK LOCUST

(Robinia pseudoacacia)

CONSIDERED AN EXCELLENT HONEY PLANT by many beekeepers, with a short but impressive blooming period, black locust attracts a variety of other bees as well as hummingbirds, bizarrely large hummingbird (sphinx) moths, and much more. The tree has sharp thorns, a shallow root system prone to suckering, and a high tolerance for drought, deicing salt, and air pollution. These traits make it hard to get rid of black locust where you don't want it, but they also make the plant highly adaptable to harsh exposed sites such as parking lot plantings. While in some places it is considered an invasive weed because it self-sows readily, it was in past times cultivated as a source of tough wood for posts and railroad ties.

EXPOSURE
Sun

SOIL MOISTURE
Dry

The honey produced from these trees is thick, watery white, and slow to granulate. Nectar production is believed to be highest in younger plants, so it may be worth cutting them back to stumps every 15 years and encouraging suckering and new growth. The sugar concentration of black locust nectar has been reported as high as 63%, with honey production of more than 170 pounds per hive. (Some of that probably comes from bees collecting aphid honeydew.)

RECOMMENDED VARIETIES

Consult federal or state noxious weed information before planting black locust. Yellow-leaved 'Frisia' is very widely available. The USDA's Natural Resources Conservation Service developed three varieties of black locust — 'Appalachia', 'Allegheny', and 'Algonquin' — for revegetation and reclamation of mining sites in the Appalachian mountain region. It is unknown whether these varieties are still widely available in the marketplace, however (we recommend checking with nurseries that specialize in plants for land restoration). We have also heard of ornamental varieties being occasionally available from commercial tree nurseries.

USES

Hedgerow

Reclaimed industrial land/ tough sites

Pollinator nesting material or caterpillar host plant

NOTABLE FLOWER VISITORS

Caterpillar host plant for many butterflies including the silver spotted skipper (*Epargyreus clarus*), golden banded skipper (*Authochton cellus*), Zarucco duskywing (*Erynnis zarucco*), funereal duskywing (*E. funeralis*), and clouded sulphur (*Colias philodice*).

BLOOM TIME	FLOWER COLOR	MAXIMUM HEIGHT
Late spring	White	80 feet (24 m)

48. BLUEBERRY
(*Vaccinium* spp.)

GLOSSY SUMMER LEAVES and fiery fall foliage make blueberry a three-season ornamental as well as an important fruit and nectar producer. It requires evenly moist but well-drained soil that is quite acidic. Cultivated varieties planted as a commercial crop typically have a short, uniform bloom period. Wild blueberries in their native habitat (and their close relatives, wild huckleberries) secrete nectar over several weeks, as individual plants have slightly different but overlapping bloom periods. Commercial beekeepers travel to Maine every year to pollinate the lowbush blueberry crop (estimated at 44,000 acres). The honey is thick and light amber in color.

EXPOSURE
Sun

SOIL MOISTURE
Dry to average, depending
on species

RECOMMENDED SPECIES OR VARIETIES

Native wild lowbush blueberry (*Vaccinium angustifolium*) occurs throughout eastern Canada, the Great Lakes, and New England, supporting enthusiastic populations of people, birds, and bees. Highbush blueberry (*V. corymbosum*), a close relative found from New England west to Michigan, has larger fruits and flowers and, like other blueberries, requires bee pollination to maximize yields. Across the South from Florida to Texas and north to North Carolina, rabbiteye blueberry (*V. virgatum*) is a common farm and garden crop. Wild relatives including cranberries (*V. macrocarpon*) and various huckleberries, bearberries, lingonberries, and deerberry are all fantastic pollinator plants.

NOTABLE FLOWER VISITORS

Small bell-shaped flowers mean the pollen and nectar are accessible only to very small bees that can climb inside or large bees with tongues long enough to reach completely in. Bumble bees will **sonicate** or buzz-pollinate the blossoms to shake off the pollen. Honey bees, carpenter bees, and short-tongued bumble bees are known to rob nectar from blueberries by biting holes in the backs of flowers. Hummingbirds visit the flowers of wild huckleberry plants. *Vaccinium* specialist bees include *Andrena bradleyi*, *A. carolina*, *Panurginus atramontensis*, *Habropoda laboriosa*, *Colletes productus*, *Osmia virga*, *Melitta americana*, and *M. eickworti*. Various blueberries are host plants for black-banded orange (*Epelis truncataria*), Canadian sphinx (*Sphinx canadensis*), and slender clearwing moth caterpillars (*Hemaris gracilis*), as well as red-spotted purple (*Limenitis arthemis*), brown elfin (*Callophrys augustinus*), and several other butterflies.

USES

Hedgerow

Pollinator nesting material or caterpillar host plant

Edible/herbal/medicinal

BLOOM TIME	FLOWER COLOR	MAXIMUM HEIGHT
Spring	White	8 feet (2.4 m)

49. BUCKWHEAT TREE

(Cliftonia monophylla)

PREFERRING WET, ACIDIC SOILS, the buckwheat tree grows in dense thickets in swamps and does not tolerate significant competition from other trees. It usually grows as a bush or small tree. Its 2- to 4-inch-long (5–10 cm) white, fragrant flower clusters produce abundant nectar. The resulting strong-flavored, almost red honey is said to be better for cooking than for direct consumption.

EXPOSURE
Sun

SOIL MOISTURE
Wet

RECOMMENDED VARIETIES

Specialty native plant nurseries are generally the sole commercial source for this wonderful tree; one cultivated variety, 'Chipolo Pink', bears pink blossoms.

NOTABLE FLOWER VISITORS

Mostly known as a high-value honey bee plant, although it no doubt supports many native bees, butterflies, and hummingbirds as well.

USES

Reforestation/shade garden

Ornamental

BLOOM TIME
Spring

FLOWER COLOR
White, pink

MAXIMUM HEIGHT
Rarely to 40 feet (12 m), although more common as a small tree or bush

50. BUTTONBUSH
(Cephalanthus occidentalis)

BUTTONBUSH OFFERS several outstanding features. It is one of the few native shrubs that provides midsummer blooms for pollinators, and one of the few that grows well in wet soils and shade. Wonderful puffball flowers and attractive foliage make it equally interesting as an ornamental plant, especially in damp locations. In restoration projects, typical applications for buttonbush include wetland revegetation and soil stabilization along streams and drainage areas. In past times buttonbush was an important source of abundant light-colored honey that supported a vibrant regional beekeeping industry along the lower Mississippi River floodplain.

EXPOSURE

Part shade to shade; tolerates sun where soils are wet

SOIL MOISTURE

Moist to wet

RECOMMENDED SPECIES OR VARIETIES

Although at least one ornamental variety of buttonbush has been developed, it doesn't offer any particular advantages over the wild species.

NOTABLE FLOWER VISITORS

Bumble bees flock to this plant, cruising the flowers for nectar and coating themselves with pollen in the process. Many classic large butterflies, hummingbird moths, and hummingbirds flutter around its blooms as well. Host plant for caterpillars of some of our largest and showiest moths, including the titan sphinx (*Aellopos titan*), the hydrangea sphinx (*Darapsa versicolor*), and the royal walnut moth (*Citheronia regalis*).

USES

Hedgerow

Reforestation/shade garden

Wetland restoration

Farm buffer/filter strip

Pollinator nesting material or caterpillar host plant

Ornamental

BLOOM TIME	FLOWER COLOR	MAXIMUM HEIGHT
Summer	White, pink	12 ft. (3.7 m)

51. CHAMISE
(Adenostoma fasciculatum)

EXPOSURE
Sun

SOIL MOISTURE
Average

ALSO CALLED GREASEWOOD because of its oily foliar secretions, chamise thrives in dry, rocky, and serpentine soils in chaparral habitats in California and parts of the Great Basin. It is often found growing in dense, solid stands or adjacent to other chaparral species such as toyon. Adapted to periodic fires, it quickly regrows from low basal crowns after burning. It makes an excellent screen or windbreak plant and is very attractive to songbirds. Clusters of small, bright white, tubular flowers arise on branch tips and are sought after by florists. The flowers, evergreen leaves, and unique, sticklike appearance (due to small leaves and flowers) make it a great (and currently underused) plant for xeriscape gardening. Chamise is an occasional source of light amber surplus honey and can help support beekeeping in areas of California that have minimal rainfall.

USES

Hedgerow

Rangeland/pasture

RECOMMENDED SPECIES OR VARIETIES

No cultivated varieties are known; however, native plant nurseries sometimes sell lower-growing (trailing) forms of the plant that originate in coastal areas.

NOTABLE FLOWER VISITORS

Attracts a variety of bee species.

BLOOM TIME	FLOWER COLOR	MAXIMUM HEIGHT
Mid-spring	White	10 feet (3 m)

52. COYOTEBRUSH
(*Baccharis* spp.)

THIS LARGE AND DIVERSE GROUP of shrubs has separate male and female plants, and only the males produce pollen, making them more attractive to most native bees. Covering extensive scrublands in parts of the Southwest, these shrubs or small trees are, interestingly, members of the aster family. Used extensively in Xerces Society pollinator hedgerow plantings in California, coyotebrush has been easy to establish with minimal care.

EXPOSURE
Sun

SOIL MOISTURE
Dry

RECOMMENDED SPECIES OR VARIETIES

Mule fat (*Baccharis salicifolia*) occurs from the Southwest all the way to South America, typically blooming from late winter through spring. Coyotebrush (*B. pilularis*) blooms from late fall through winter and is well adapted in coastal and northern areas including cool (but not cold) parts of California and the Pacific Northwest. Groundsel (*B. halimifolia*) blooms in fall in the eastern United States.

NOTABLE FLOWER VISITORS

The very late bloom period provides a key resource for honey bees and for prehibernation or early-emerging wild bees. Coyotebrush is closely related to the famous Brazilian "green baccharis" plant (*Baccharis dracuncufolia*), which is the source of high-value medicinal **propolis** (the plant resin collected by honey bees for hive construction). Coyotebrush is a caterpillar host plant for a few moth and butterfly species, but little is known about them.

USES

Hedgerow

Rangeland/pasture

Pollinator nesting material or caterpillar host plant

BLOOM TIME
Variable, depending on species and location, with coyotebrush blooming in fall in some areas and/or in spring in others

FLOWER COLOR
White

MAXIMUM HEIGHT
8 feet (2.4 m)

53. FALSE INDIGO, LEADPLANT

(*Amorpha* spp.)

THE COMBINATION OF DELICATE compound leaves and purple flower spikes makes false indigo an impressively showy landscape plant when it is in full bloom. Commonly found along rivers, ponds, and in semiwet soils, it is well adapted to drainage areas and, if watered during the first few years after planting, can easily adapt to dry upland sites as well. Note that in some locations it is considered weedy and thus should not be introduced beyond its native range. Short, slow-growing leadplant is not aggressive and is well adapted to dry, poor-quality soils.

EXPOSURE
Sun

SOIL MOISTURE
Dry to wet

Various debates swirl around the origin of leadplant's name. According to some sources, the name comes from the pewter-colored leaves, while others claim it references early folk wisdom about where to dig for galena deposits. Regardless of the truth, the plant is a good candidate for revegetating dry abandoned quarries, gravel pits, and strip-mining sites in the Midwest.

RECOMMENDED SPECIES OR VARIETIES

The two most common species are the larger false indigo (*Amorpha fruticosa*) and leadplant (*A. canescens*).

NOTABLE FLOWER VISITORS

Attracts bumble bees (*Bombus* spp.), leafcutter bees (*Megachile* spp.), polyester bees (*Colletes* spp.), and various sweat bees. The mining bee (*Andrena quintilis*) is thought to be a specialist pollen collector of leadplant. Both species are host plants for caterpillars of clouded sulphur (*Colias philodice*), dogface sulphur (*Zerene cesonia*), gray hairstreak (*Strymon melinus*), hoary edge (*Achalarus lyciades*), marine blue (*Lepototes marina*), and silver-spotted skipper (*Epargyreus clarus*) butterflies, as well as the Io moth (*Automeris io*).

USES

Hedgerow

Wildflower meadow/ prairie restoration

Reclaimed industrial land/tough sites

Rangeland/pasture

Farm buffer/filter strip

Pollinator nesting material or caterpillar host plant

Ornamental

BLOOM TIME
Summer

FLOWER COLOR
Purple

MAXIMUM HEIGHT
10 feet (3 m)

54.
GOLDEN CURRANT

(Ribes aureum)

I N EARLY SPRING, this western plant produces an abundance of bright yellow flowers notable for their clove or vanilla fragrance. Various early-season bee species visit, helping to produce red berries that provide food for humans and wildlife late in summer. (If you want to eat the berries, expect to add a sweetener. Try nibbling the flowers; they are edible as well.) The plant thrives in dry, exposed locations, spreads by suckering, and is appropriate for revegetation of many sites, as well as for hedgerow use across the inland West.

EXPOSURE
Sun to part shade

SOIL MOISTURE
Dry to average

RECOMMENDED VARIETIES

Several recognized subspecies exist across the West. *Ribes aureum* subsp. *aureum* and subsp. *villosum* occur inland; *R. aureum* subsp. *gracillimum* is found in California's coastal mountain ranges.

NOTABLE FLOWER VISITORS

Attracts various early-season wild bee species as well as hummingbirds and butterflies. It was once thought that the blossoms' deep nectar tubes were inaccessible to honey bees, but these insects are common visitors when the plant is in full bloom.

USES

Hedgerow

Rangeland/pasture

Ornamental

Edible/herbal/ medicinal

BLOOM TIME	FLOWER COLOR	MAXIMUM HEIGHT
Mid-spring	Yellow	10 feet (3 m)

55. INKBERRY
(Ilex glabra)

BEEKEEPER LORE IS RICH with anecdotes about honey bees ignoring other floral sources when inkberry is in bloom. Whether true or not, it does supply light-colored nongranulating honey with yields reportedly approaching 300 pounds per hive. Also known as gallberry, this native holly is widespread in the eastern United States and Canada but most common in the Southeast, where it has a well-deserved reputation for productivity among beekeepers. Plants bloom for roughly a month, producing more than 100 pounds of surplus honey per hive in optimal locations. Inkberry is a good bee-scaping choice for moist lowland soils. In addition to supporting bees, the berries of this shrub or small tree are an important source of winter food for birds. Ornamental gardeners value it for its small, glossy evergreen leaves and its berries that turn black like buttonbush berries, and its ability to thrive in part shade and in moist soils.

EXPOSURE
Part shade

SOIL MOISTURE
Moist to wet

Closely related American holly (*Ilex opaca*, below, right) is also a good forage plant for honey bees; it tolerates a slightly greater range of conditions, thriving in much of the southeastern United States.

USES

Hedgerow

Reforestation/shade garden

Pollinator nesting material or caterpillar host plant

NOTABLE FLOWER VISITORS

Attracts bees and the occasional hummingbird. *Colletes banksi* is a specialist bee. Inkberry is a caterpillar host plant for holly azure (*Celastrina idella*) and Henry's elfin (*Callophrys henrici*) butterflies, as well as the pawpaw sphinx moth (*Dolba hyloeus*).

BLOOM TIME	FLOWER COLOR	MAXIMUM HEIGHT
Summer	White	9 feet (2 m)

56. MADRONE
(*Arbutus* spp.)

ADAPTED TO HUMID COASTAL SITES as well as dry foothills and canyon areas, madrone serves as a good erosion-control species and readily thrives in areas prone to frequent disturbance, including periodic burning. Shiny terra-cotta–colored bark, and the potential to reach towering heights on slender trunks, makes these graceful trees a wonderful ornamental addition to landscapes (although they can be notoriously difficult to establish as transplants). Clusters of white, bell-shaped flowers in mid-spring develop into red berries that provide food for birds, deer, and other wildlife. Madrone is also a prolific nectar producer with an average sugar concentration of 15%.

EXPOSURE
Sun to part shade

SOIL MOISTURE
Dry

RECOMMENDED SPECIES OR VARIETIES

The most common species is Pacific madrone (*Arbutus menziesii*).

USES

Reforestation/shade garden

Pollinator nesting material or caterpillar host plant

Ornamental

NOTABLE FLOWER VISITORS

In the Northwest, numerous hummingbirds can often be seen feeding on the same tree. Host plant for caterpillars of ceanothus silkmoth (*Hyalophora euryalus*) and Mendocino saturnia moth (*Saturnia mendocino*), two of the larger and showier moths found on the West Coast, and the brown elfin butterfly (*Callophrys augustinus*).

BLOOM TIME
Mid-spring

FLOWER COLOR
White

MAXIMUM HEIGHT
100 feet

57. MAGNOLIA
(*Magnolia* spp.)

CONSIDERED VERY PRIMITIVE in their physical structure, magnolia flowers represent an early evolutionary precursor to many other types of flowering plants that exist today. There are roughly a dozen species in this genus that are native to North America, with most found in the southeastern United States, and a few introduced species that are now found elsewhere. Note that some magnolias are deciduous while others are evergreen. All grow best in rich, slightly acidic soil.

EXPOSURE
Sun to shade

SOIL MOISTURE
Average

RECOMMENDED SPECIES

Some of the more common, widely distributed, and easy-to-acquire species include southern magnolia (*Magnolia grandiflora*), big-leaf magnolia (*M. macrophylla*), and sweet bay (*M. virginiana*).

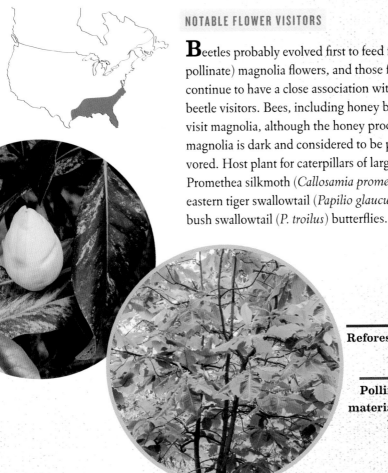

NOTABLE FLOWER VISITORS

Beetles probably evolved first to feed from (and pollinate) magnolia flowers, and those flowers today continue to have a close association with various beetle visitors. Bees, including honey bees, will also visit magnolia, although the honey produced from magnolia is dark and considered to be poorly flavored. Host plant for caterpillars of large and showy Promethea silkmoth (*Callosamia promethean*) and eastern tiger swallowtail (*Papilio glaucus*) and spicebush swallowtail (*P. troilus*) butterflies.

USES

Hedgerow

Reforestation/shade garden

Pollinator nesting material or caterpillar host plant

Ornamental

BLOOM TIME
Spring

FLOWER COLOR
White

MAXIMUM HEIGHT
40 feet (12 m)

58. MANZANITA

(*Arctostaphylos* spp.)

RELATED TO BLUEBERRIES, manzanitas make up a huge group of evergreen flowering trees and shrubs, with the greatest diversity found in California. Kinnikinnick or bearberry, a low-growing species found throughout most of the United States and Canada, is an excellent pollinator plant and a great ground cover. Some manzanitas are said to have so much nectar that it can easily be shaken from flowers on warm days. In parts of California an occasional surplus honey crop has been reported from these plants, perhaps resulting from the long bloom time (more than a month) of some species — and from the high sugar concentration of manzanita nectar, reported to range from 16 to 50%. Manzanita honey has been described as light in color but slightly bitter.

EXPOSURE
Sun

SOIL MOISTURE
Average

RECOMMENDED SPECIES

Kinnikinnick or bearberry (*Arctostaphylos uva-ursi*) is readily available and good for northern and mountain regions. In the West, common manzanita (*A. manzanita*), greenleaf manzanita (*A. patula*), bigberry manzanita (*A. glauca*), and Pajaro manzanita (*A. pajaroensis*) appear occasionally at specialty native plant nurseries, especially in California.

USES

Hedgerow

Reforestation/shade garden

Pollinator nesting material or caterpillar host plant

Ornamental

NOTABLE FLOWER VISITORS

Attracts hummingbirds, bumble bees, honey bees, mason bees, and other common bee species. Various members of this genus are also host plants for caterpillars of the showy Mendocino saturnia moth (*Saturnia mendocino*) and ceanothus silkmoth (*Hyalophora euryalus*), as well as hoary elfin (*Callophrys polios*) and black-banded orange (*Epelis truncataria*) butterflies.

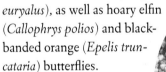

BLOOM TIME	FLOWER COLOR	MAXIMUM HEIGHT
Spring	White, pink	20 feet (6 m)

59. MESQUITE
(*Prosopis* spp.)

TOUGH, DROUGHT-TOLERANT TREES, mesquites thrive in hot, sunny environments where both humans and livestock value their shade in summer months. They are native to arid and tropical regions and one of the most important honey sources in Texas, the Southwest, and Mexico. The light amber honey regularly shows up even in large grocery stores, commanding decent prices. Honey crops of more than 200 pounds have often been reported from mesquite trees, although nectar flow can vary depending on weather conditions. It is believed that the best nectar flows occur after a wet spring followed by very hot summer weather. Like acacias (which mesquite resembles), these plants also support birds with seeds and nesting sites.

EXPOSURE
Sun

SOIL MOISTURE
Dry

RECOMMENDED SPECIES OR VARIETIES

The most common mesquite in Texas and the Southwest is tornillo (*Prosopis odorata*).

NOTABLE FLOWER VISITORS

Attracts bees and butterflies seeking nectar. Host plant for caterpillars of tailed orange (*Eurema proterpia*), marine blue (*Leptotes marina*), ceraunus blue (*Hemiargus ceraunus*), and Reakirt's blue (*Echinargus isola*) butterflies.

USES

Hedgerow

Rangeland/pasture

Pollinator nesting material or caterpillar host plant

BLOOM TIME	FLOWER COLOR	MAXIMUM HEIGHT
Spring through summer	Yellow	36 feet (11 m)

60. OCEAN SPRAY

(Holodiscus discolor)

THIS VERY ADAPTABLE WESTERN PLANT is equally at home on dry, rocky coastal cliffs, in cool forest understories, and on mountain slopes. Its thick foliage provides cover for numerous birds, small mammals, and even tree frogs. Equally valuable, ocean spray's white masses of cascading flowers appear in midsummer, when most other shrubs and wildflowers have finished for the year. It attracts huge numbers of beneficial insects and is a staple of Xerces Society's pollinator hedgerows in the Pacific Northwest.

EXPOSURE
Sun to shade

SOIL MOISTURE
Average to dry

RECOMMENDED VARIETIES

Although a cultivated variety of ocean spray has apparently been developed in Europe, the Xerces Society uses and recommends the standard species for habitat restoration.

USES

Hedgerow

Reforestation/shade garden

Pollinator nesting material or caterpillar host plant

Ornamental

NOTABLE FLOWER VISITORS

Ocean spray hums with activity when in full bloom, drawing all manner of wasps, syrphid flies, and wild bees, large and small. Caterpillar host plant for Lorquin's admiral (*Limenitis lorquini*), pale tiger swallowtail (*Papilio eurymedon*), and spring azure (*Celastrina ladon*) butterflies.

BLOOM TIME	FLOWER COLOR	MAXIMUM HEIGHT
Summer	White	20 feet (6 m)

61. OREGON GRAPE
(Mahonia aquifolium)

A HOLLYLIKE EVERGREEN SHRUB, Oregon grape has bright yellow bell-shaped blossoms in spring that give way to clusters of edible blue fruit later in the year. All of these handsome features, along with its adaptability to many sites and conditions, have made Oregon grape a popular landscaping shrub. Also called holly-leaved barberry, this species exists in two widely separate ranges: an eastern population in the Great Lakes region, and a western population in the Pacific Northwest.

EXPOSURE
Shade to part shade

SOIL MOISTURE
Wet to dry

RECOMMENDED SPECIES OR VARIETIES

Numerous cultivated varieties of Oregon grape are now available, including dwarf varieties popular for ornamental landscapes. At Xerces, we prefer the standard wild type for our habitat restoration projects.

USES

Hedgerow

Reforestation/shade garden

Ornamental

Edible/herbal/medicinal

NOTABLE FLOWER VISITORS

Attracts bumble bees (*Bombus* spp.) most of all, although mason bees (*Osmia* spp.) and hummingbirds — especially Anna's hummingbird (*Calypte anna*) — are also common flower visitors.

BLOOM TIME	FLOWER COLOR	MAXIMUM HEIGHT
Spring	Yellow	12 feet (3.6 m)

62.
RABBITBRUSH
(Chrysothamnus spp.)

BLOOMING LATE IN THE YEAR, even after unexpected frosts have killed other flowering plants, makes rabbitbrush an important nectar source in desert regions where some butterflies may remain active throughout winter. Usually found growing with sagebrush, these are tough, drought-tolerant plants common in arid regions of western North America. As well as being useful rangeland plants, their abundant blooms add a cheerful late-season splash of yellow color to xeriscape gardens. The dark-colored, foul-smelling honey produced from these plants is not popular for human consumption, but it can be an important source of overwintering food for honey bees in western states.

EXPOSURE
Sun

SOIL MOISTURE
Dry

LEFT Honey bee on rabbitbrush

RECOMMENDED SPECIES OR VARIETIES

Yellow rabbitbrush (*Chrysothamnus viscidiflorus*), Greene's rabbitbrush (*C. greenei*).

NOTABLE FLOWER VISITORS

Host plant for caterpillars of sagebrush checkerspot butterfly (*Chlosyne acastus*).

USES

Wildflower meadow/ prairie restoration

Rangeland/pasture

Pollinator nesting material or caterpillar host plant

Ornamental

BLOOM TIME	FLOWER COLOR	MAXIMUM HEIGHT
Fall	Yellow	5 feet (1.5 m)

63.

REDBUD

(Cercis spp.)

BRILLIANT MAGENTA FLOWERS open along the bare stems of redbud in early spring before the leaves emerge, a striking effect that rivals any highly bred ornamental and has endeared this small native tree to many home gardeners. The very early bloom of this plant makes it an important source of spring pollen for many bees. Redbud grows best in rich, moist soils, sheltered locations, and moderate temperatures. It does not thrive in extremely cold climates or high elevations.

EXPOSURE
Sun to shade

SOIL MOISTURE
Average

RECOMMENDED SPECIES OR VARIETIES

In the West, California redbud (*Cercis orbiculata*) adapts to most locations and will even survive in desert areas if watered during the first few years after planting. In the East and Midwest, the Eastern redbud (*C. canadensis*) is readily available from both native and ornamental nurseries.

NOTABLE FLOWER VISITORS

An important source of early spring pollen for many bees, including some unusual species such as the southeastern blueberry bee (*Habropoda laboriosa*), which Xerces staff have observed visiting redbud flowers in northern Florida. Even in midsummer, after the flowers are gone, redbud continues to support leafcutter bees (*Megachile* spp.) as a preferred source of leaf pieces for nesting material. Although not tremendously attractive to butterflies, redbud is a caterpillar host plant for the Henry's elfin butterfly (*Callophrys henrici*).

USES

Hedgerow

Reforestation/shade garden

Pollinator nesting material or caterpillar host plant

Ornamental

ABOVE
Brown-belted bumble bee on redbud

BLOOM TIME
Early spring

FLOWER COLOR
Magenta

MAXIMUM HEIGHT
30 feet (9 m)

64. RHODODENDRON
(*Rhododendron* spp.)

MORE THAN 1,000 SPECIES of these magnificent, bold-flowering, glossy-leaved plants are found across the world. Some of the greatest species diversity occurs in Asia and the Appalachian Mountains (although other locations have their native share of these showy flowering shrubs). For years, taxonomists have grappled with the categorization of these plants, frequently revising and regrouping both azaleas and rhododendrons. Although exceptions exist, most of these species are evergreen, grow in moderate temperate regions, and prefer moist, acidic soils.

EXPOSURE
Sun to shade

SOIL MOISTURE
Average to moist

RECOMMENDED SPECIES OR VARIETIES

In the East, smooth azalea (*Rhododendron arborescens*), piedmont rhododendron (*R. minus*), and early azalea (*R. prinophyllum*) are all readily available from nurseries. In the West, look for native Pacific rhododendron (*R. macrophyllum*) and the Cascade azalea (*R. albiflorum*). Numerous ornamental varieties are also available, including selections of native species such as great laurel (*R. maximum*) and swamp azalea (*R. viscosum*), as well as hybrids.

USES

Hedgerow

Reforestation/shade garden

Pollinator nesting material or caterpillar host plant

Ornamental

NOTABLE FLOWER VISITORS

Attracts bumble bees, which eagerly collect nectar with no obvious ill effects. Honey bees, however, have occasionally been documented to die from rhododendron nectar, due to the presence of the chemical andromedotoxin. This chemical, once further concentrated in honey, can also be lethal to humans. This phenomenon has been described since early times, such as a famous mass poisoning of the Persian army in 401 BCE after soldiers supposedly consumed rhododendron honey. Rhododendron species vary significantly in their relative toxicity. *Andrena cornelli* is a specialist bee of *Rhododendron*. Host plant for caterpillars of several large and ornate butterflies, including the green comma (*Polygonia faunus*), gray comma (*P. progne*), and hoary comma (*P. gracilis*).

BLOOM TIME	FLOWER COLOR	MAXIMUM HEIGHT
Spring	Many hues	20 feet (6 m)

65. ROSE
(Rosa spp.)

OPEN-FLOWERED NATIVE ROSES are dependable bee plants: in addition to offering attractive flowers and edible rose hips, they are generally easy to grow and tough. Most modern hybrids, on the other hand, offer little or nothing to pollinators — except those with flat single-petaled flowers that resemble wild roses. Most roses need average to dry soil, but a few such as swamp rose (*Rosa palustris*) tolerate wet roots. Native roses work well as hedgerow and hedgerow understory plants (depending on their mature size). They incorporate equally well into shelterbelts, riparian buffer areas, and shrubby pastures. Most roses produce little nectar, so pollen-collecting bees are the primary visitors.

EXPOSURE
Sun to part shade

SOIL MOISTURE
Wet to dry

RECOMMENDED SPECIES OR VARIETIES

In the West, Woods' rose (*Rosa woodsii*) and Nootka rose (*R. nutkana*) are common, easy-to-grow species. In the Central Plains, prairie rose (*R. arkansana*) is well adapted to most sites. In the Midwest and East, Virginia rose (*R. virginiana*), swamp rose (*R. palustris*), and Carolina rose (*R. carolina*) are widely distributed and readily available options.

USES

Hedgerow

Wildflower meadow/ prairie restoration

Farm buffer/filter strip

Pollinator nesting material or caterpillar host plant

Ornamental

Edible/herbal/medicinal

NOTABLE FLOWER VISITORS

Pollen-collecting bees are the primary visitors, including the specialist bee *Eucera rosae*. Leafcutter bees (*Megachile* spp.) harvest bits of rose leaves to use in nest construction; this does no significant damage to plants. Host plant for the bizarre-looking caterpillars of the stinging rose caterpillar moth (*Parasa indetermina*). As the name suggests, these brightly colored caterpillars (which resemble sea anemones!) cause stinging and skin burns when touched. As adults, these large moths resemble curled green leaves.

RIGHT Leafcutter bee on rose

BLOOM TIME	FLOWER COLOR	MAXIMUM HEIGHT
Late spring to summer	White, yellow, red, pink	12 feet (3.6 m)

66.

SAW PALMETTO

(Serenoa repens)

A LOW-GROWING SCRUB PALM, saw palmetto is ubiquitous in thickets in sandy soils across Florida and Gulf Coast woodlands. Although the plant is not typically available from commercial nurseries, it is worth recognizing its value and prioritizing its conservation as part of the woodland plant community. Despite their small size, saw palmetto plants are extremely long lived (possibly hundreds of years) and able to resprout following forest fires, seeming to flower more abundantly on new growth. Among beekeepers, saw palmetto is famous for its abundant spring nectar flow. It produces thick, delicious, lemon-yellow honey that some argue is the best produced in Florida. Beekeepers regularly report 50- to 80-pound honey surpluses per colony from this plant.

EXPOSURE
Part shade

SOIL MOISTURE
Well-drained or seasonally
waterlogged soils

RECOMMENDED SPECIES OR VARIETIES

Various other palmettos are found in subtropical and tropical regions, many of them excellent pollinator plants, but the common wild saw palmetto is among the best.

NOTABLE FLOWER VISITORS

Attracts honey bees. Host plant for caterpillars of palmetto skipper butterfly (*Euphyes arpa*).

USES

Reforestation/shade garden

Wetland restoration

Pollinator nesting material
or caterpillar host plant

Edible/herbal/medicinal

BLOOM TIME	FLOWER COLOR	MAXIMUM HEIGHT
Spring to summer	White	10 feet (3 m)

67.
SERVICEBERRY
(*Amelanchier* spp.)

ALSO KNOWN AS SHADBUSH OR SHADBLOW, several serviceberry species are planted as ornamentals and as fruit trees. Many species are fire dependent in the wild, growing most abundantly in forest areas recently cleared by burning. All are excellent pollinator plants, often among the earliest blooming plants wherever they grow — with their clouds of delicate white flowers conspicuously standing out in forests of otherwise bare spring trees.

EXPOSURE	SOIL MOISTURE
Sun to part shade	Average

RECOMMENDED SPECIES

In the West, Saskatoon serviceberry (*Amelanchier alnifolia*) is common and widely available. In the East, common serviceberry (*A. arborea*) and Canadian serviceberry (*A. canadensis*) are both widely available from native plant vendors and regular nurseries.

USES

Hedgerow

Reforestation/shade garden

Pollinator nesting material or caterpillar host plant

Ornamental

NOTABLE FLOWER VISITORS

The spring blossoms of serviceberry attract various native bee species and honey bees. Host plant for caterpillars of various large, showy butterflies including Weidemeyer's admiral (*Limenitis weidemeyerii*), western swallowtail (*Papilio zelicaon*), pale swallowtail (*P. eurymedon*), and two-tailed swallowtail (*P. multicaudata*).

BLOOM TIME	FLOWER COLOR	MAXIMUM HEIGHT
Early spring	White	40 feet (12 m)

68. SOURWOOD
(Oxydendrum arboreum)

THESE HANDSOME PYRAMID-SHAPED TREES thrive in moist, peaty, acidic soil. In fall they produce striking scarlet-red foliage that rivals the color of any maple. This beautiful fall show endears sourwood to home gardeners. In the wild it grows in mixed-oak forests and typically remains small under the competition of a larger oak tree canopy; in open areas without shade or competition it can grow much larger. Songs and other folklore extol its value as a honey plant, a reputation that is well earned. Sourwood is a prolific nectar plant, bearing long, slender clusters of white bell-shaped flowers. Its honey is famous among beekeepers and foodies alike, with a light, lavender hue, a thick texture, and a slight hint of maple flavor. It is very slow to granulate. Surpluses of more than 70 pounds per colony have been reported.

EXPOSURE
Sun to shade

SOIL MOISTURE
Average

RECOMMENDED SPECIES

Several ornamental cultivars are available, including 'Chameleon' and 'Mt. Charm'. These have been selected mostly for fall color and may not represent the best selections for pollinators.

NOTABLE FLOWER VISITORS

Attracts honey bees, bumble bees, carpenter bees, leafcutter bees, mason bees, and resin bees.

USES

Reforestation/shade garden

Ornamental

BLOOM TIME	FLOWER COLOR	MAXIMUM HEIGHT
Early summer	White	80 feet (24 m)

69. STEEPLEBUSH, MEADOWSWEET

(Spiraea spp.)

SPIREA SPECIES TEND to prefer fertile, slightly wet soils such as ditches, stream banks, and the edges of grassed waterways on farms. Initially slow growing, once established the plants can hold their own against many invasive species such as reed canary grass and Himalayan blackberry. Nearly all spireas integrate wonderfully into hedgerows and complement diverse shrub plantings with their large spearhead-shaped flower clusters and very long-lasting blooms.

EXPOSURE
Sun

SOIL MOISTURE
Average to wet

RECOMMENDED SPECIES

Steeplebush (*Spiraea tomentosa*) and white meadowsweet (*S. alba*) are two widely distributed eastern species with long bloom periods. In the west, Douglas spirea (*S. douglasii*) is widely available from nurseries and establishes easily anywhere with sufficient moisture.

NOTABLE FLOWER VISITORS

Attracts a wide variety of beneficial insects, especially small flies and wasps, also butterflies and bumble bees.

USES

Hedgerow

Reforestation/shade garden

Wetland restoration

Farm buffer/filter strip

Ornamental

BLOOM TIME	FLOWER COLOR	MAXIMUM HEIGHT
Late spring to early fall	White, pink, purple	4 feet (1.2 m)

70. TOYON
(Heteromeles arbutifolia)

ONCE A MAJOR COMPONENT of California's chaparral ecosystem, toyon has dark green leathery evergreen leaves rising from multiple stems and long-lasting red berries that are consumed by birds (and frequently used in holiday wreaths). It is ideal as a specimen shrub or as a screen when planted as a hedge. Its ornamental qualities and ability to survive drought, below-freezing temperatures, and generally tough sites have made it increasingly common as a landscape plant outside California. The average sugar concentration of toyon nectar has been reported as 44%. It produces a thick amber honey that crystallizes easily. The hollylike leaves and berries of toyon are supposedly the namesake for the city of Hollywood — and not holly itself.

EXPOSURE
Sun

SOIL MOISTURE
Dry

RECOMMENDED SPECIES OR VARIETIES

Toyon can vary in mature size; somewhat distinct local populations occur as small shrubs while others grow into a small tree. When buying at a nursery that sells native plants, ask what their experience is with the mature size and shape of the toyon they sell.

USES

Hedgerow

Reclaimed industrial land/ tough sites

Rangeland/pasture

Pollinator nesting material or caterpillar host plant

Ornamental

NOTABLE FLOWER VISITORS

A tough, all-purpose bee plant, supporting native bees and honey bees alike. Caterpillar host plant for the echo azure butterfly (*Celastrina echo*).

BLOOM TIME	FLOWER COLOR	MAXIMUM HEIGHT
Spring to early summer	White	30 feet (9 m)

71. TULIP TREE
(*Liriodendron tulipifera*)

ITS BIG SHOWY FLOWERS make tulip tree an attractive addition to a residential landscape, and many cultivated varieties are now available from nurseries. The tall trees produce large quantities of nectar — in some cases they literally drip nectar — that attracts honey bees and wild pollinators alike.

EXPOSURE
Sun

SOIL MOISTURE
Average to wet

RECOMMENDED VARIETIES

Choose plants with bright yellow flowers, because those with greener tints may be less attractive to pollinators.

USES

Reforestation/shade garden

Pollinator nesting material or caterpillar host plant

Ornamental

NOTABLE FLOWER VISITORS

Attracts honey bees and native bees. Caterpillar host plant for the eastern tiger swallowtail (*Papilio glaucus*) and the spicebush swallowtail (*P. troilus*) butterflies.

BLOOM TIME	FLOWER COLOR	MAXIMUM HEIGHT
Mid-spring	Yellow	120+ feet (36.5 m)

72. TUPELO

(*Nyssa* spp.)

EXPOSURE	SOIL MOISTURE
Sun	Average to wet

F AMOUS AMONG BEEKEEPERS, water tupelo produces a very light, mild, noncrystallizing honey that commands a high market price. The more swamp-adapted species are the best honey producers, supporting a million-dollar specialty honey industry in Florida. Florida beekeepers even place hives on floating platforms along river swamps to take advantage of the tupelo bloom. Historically, thousands of barrels of honey were produced in the Apalachicola area annually. Tupelo trees are most at home in swamps and seasonally wet lowland soils. While the flowers are not particularly showy, scarlet autumn foliage has made this increasingly popular as an ornamental landscape tree.

RECOMMENDED SPECIES

W ater tupelo (*Nyssa aquatica*) and white tupelo (*N. ogeche*) are the most important and prolific honey producers. Black gum (*N. sylvatica*) is more widely distributed, and more adaptable to drier upland soils.

NOTABLE FLOWER VISITORS

A ttracts honey bees.

USES

Reforestation/shade garden

Ornamental

BLOOM TIME	FLOWER COLOR	MAXIMUM HEIGHT
Mid-spring	Green	50 feet (15 m)

73. WILD LILAC

(*Ceanothus* spp.)

THEIR WONDERFUL ORNAMENTAL QUALITIES (some produce electric-blue flowers), fragrant blooms, and leathery leaves make wild lilacs excellent plants for the home landscape and farm hedgerow alike. An enormous diversity of bees and other pollinators visit wild lilacs, which in the West are commonly referred to by their genus name, *Ceanothus*. Most species are slow growing, difficult to establish from seed, and readily eaten by deer.

EXPOSURE
Sun to part shade

SOIL MOISTURE
Average

RECOMMENDED SPECIES

In the West, many native species are readily available from nurseries, especially in California. The Xerces Society often plants buckbrush (*Ceanothus cuneatus*) in pollinator hedgerows. In the Midwest and New England, the white-flowered New Jersey tea (*C. americanus*) is a common and adaptable species.

NOTABLE FLOWER VISITORS

Attracts bees, syrphid and tachinid flies, mud daubers, spider wasps, sand wasps, and many butterflies. *Pseudopanurgus pauper* is a specialist bee. Host plant for caterpillars of California hairstreak (*Satyrium californica*), hedgerow hairstreak (*S. saepium*), California tortoiseshell (*Nymphalis californica*), spring azure (*Celastrina ladon*), summer azure (*C. neglecta*), echo blue (*C. echo*), mottled duskywing (*Erynnis martialis*), pacuvius duskywing (*E. pacuvius*), and western green hairstreak (*Callophrys affinis*) butterflies as well as white-streaked saturnia (*Saturnia albofasciata*) and ceanothus silk (*Hyalophora euryalus*) moths.

USES

Hedgerow

Rangeland/pasture

Pollinator nesting material or caterpillar host plant

Ornamental

BLOOM TIME	FLOWER COLOR	MAXIMUM HEIGHT
Spring to summer	White, blue, pink	12 feet (3.6 m)

74. WILLOW

(*Salix* spp.)

A N IMPORTANT SPRING FOOD SOURCE for bees, willows offer the first pollen available in many areas. To take advantage of this, beekeepers sometimes plant willow thickets around apiaries to provide both a windbreak and an early-season source of nutrition. Willows have separate male and female plants, so plant males (the only ones that produce pollen) if bee forage is the primary goal. Both male and female willows can produce nectar, with the sugar concentration in nectar reported as high as 60% in some species. Willows are easily propagated by cuttings that are simply planted in the ground and regularly watered. They also readily resprout from **coppice cutting** (cutting them back to stumps), creating dense, multistem thickets.

EXPOSURE
Sun to part shade

SOIL MOISTURE
Wet to average

RECOMMENDED SPECIES

Pussy willows are among the better species, especially the native pussy willow (*Salix discolor*) and the nonnative *S. caprea* and *S. cinerea*. Horticultural hybrids, including most weeping willows, are of little value to pollinators.

USES

Hedgerow

Wetland restoration

Farm buffer/filter strip

**Pollinator nesting material
or caterpillar host plant**

Ornamental

NOTABLE FLOWER VISITORS

Attracts many specialist mining bee species, including *Andrena andrenoides*, *A. bisalicis*, *A. erythrogaster*, *A. fenningeri*, *A. illinoiensis*, *A. mariae*, *A. salictaria*, and *A. sigmundi*. Host plant for caterpillars of many butterflies, including eastern tiger swallowtail (*Papilio glaucus*), western tiger swallowtail (*P. rutulus*), mourning cloak (*Nymphalis antiopa*), Compton's tortoiseshell (*N. vaualbum*), Lorquin's admiral (*Limenitis lorquini*), Weidemeyer's admiral (*L.s weidemeyerii*), white admiral (*L. arthemis*), viceroy (*L. archippus*), Acadian hairstreak (*Satyrium arcadica*), California hairstreak (*S. californica*), Sylvan hairstreak (*S. sylvinus*), and dreamy duskywing (*Erynnis icelus*).

Also host plant for caterpillars of many large and dramatically patterned moths including black-waved flannel (*Lagoa crispata*), Cynthia (*Samia cynthia*), imperial (*Eacles imperialis*), Io (*Automeris io*), polyphemus (*Antheraea polyphemus*), promethean (*Callosamia promethea*), cecropia (*Hyalophora cecropia*), elm sphinx (*Ceratomia amyntor*), twin-spotted sphinx (*Smerinthus jamaicensis*), blinded sphinx (*Paonias excaecata*), and modest sphinx (*Pachysphinx modesta*).

BLOOM TIME	FLOWER COLOR	MAXIMUM HEIGHT
Early spring	White, yellow	140 feet (43 m) (*S. nigra*)

75. YERBA SANTA

(*Eriodictyon* spp.)

BEES, BUTTERFLIES, AND HUMMINGBIRDS all flock to yerba santa blossoms, and honey bees produce a spicy, amber-colored honey from its nectar. Despite its magnificent pollinator value, these oily plants are unlikely to win praise for their unpleasant smell and sticky, resin-covered leaves and stems, which are flammable and often covered in fungus. Like other natives of the California chaparral, yerba santa thrives in recently burned areas and quickly regrows in shrubby clonal colonies, resprouting from extensive rhizomes.

EXPOSURE
Sun

SOIL MOISTURE
Dry

RECOMMENDED SPECIES

Most species are limited to California and may be available only from specialty native plant nurseries. Look for narrowleaf yerba santa (*Eriodictyon angustifolium*) and the showier California mountain balm (*E. californicum*).

NOTABLE FLOWER VISITORS

Attracts honey bees, countless species of wild bees, hummingbirds, and butterflies. Host plant for caterpillars of the large and showy pale swallowtail butterfly (*Papilio eurymedon*).

USES

Hedgerow

Rangeland/pasture

Pollinator nesting material or caterpillar host plant

Medicinal

ABOVE
Variable checkerspot butterfly on yerba santa

BLOOM TIME	FLOWER COLOR	MAXIMUM HEIGHT
Spring	Purple, white	8 feet (2.4 m)

Introduced Trees and Shrubs

Floral and foliage rewards from introduced woody plants may sustain numerous pollinator populations. Introduced trees and shrubs that are nonweedy can be combined with native trees and shrubs in gardens and hedgerows for many benefits. Here, we present a few of our favorite introduced trees and shrubs for pollinators.

76. ORANGE
(Citrus sinensis)

ORANGE BLOSSOMS and other *Citrus* blooms were formerly important for honey production, especially in Florida. The increase of pesticide use on citrus farms and the popularity of seedless orange varieties (which don't require bee pollination) have led to a decline in orange blossom honey. In some regions orange growers have actively worked to discourage nearby beekeeping, since pollen contamination by bees can cause seeds to develop in seedless oranges. When orange blossom honey was common, however, single colonies could produce more than 100 pounds of surplus honey in a good season. This honey is still much sought after: thick, white, and with a wonderful floral aroma. The average sugar concentration in the nectar of some orange species has been reported at 25%.

EXPOSURE
Sun

SOIL MOISTURE
Well draining

RECOMMENDED SPECIES OR VARIETIES

In addition to *Citrus sinensis* (and the many varieties of orange), kumquats (*C. japonica*) and other related fruits produce an abundance of bee-attracting flowers. One such related species valued for its fruit and slightly greater cold tolerance is the Satsuma orange (*C. unshiu*).

NOTABLE FLOWER VISITORS

Attracts honey bees; caterpillar host plant for sickle-winged skipper (*Eantis tamenund*), Thoas swallowtail (*Papilio thoas*), eastern giant swallowtail (*P. cresphontes*), and broad-banded swallowtail (*P. astyalus*) butterflies.

USES

Hedgerow

Pollinator nesting material or caterpillar host plant

Ornamental

Edible/herbal/medicinal

BLOOM TIME
Early spring, early winter

FLOWER COLOR
White

MAXIMUM HEIGHT
50 feet (15 m)

77. PLUM, CHERRY, ALMOND, PEACH

(*Prunus* spp.)

THE *PRUNUS* GENUS includes dozens of native and introduced spring-blooming trees and shrubs. Nearly all are high-value pollinator plants, supporting early-season bee species such as mining bees and mason bees. Many of these plants, especially the smaller shrubby wild plum species, respond well to **coppice cutting**. This technique involves periodically cutting the main trunk back to the ground to encourage suckering and a resulting denser, shorter shrublike form useful for hedgerows. The average sugar concentration in the nectar of some plum and cherry species reportedly ranges from 15 to 40%.

EXPOSURE
Sun to part shade

SOIL MOISTURE
Average

RECOMMENDED SPECIES

In addition to cultivated species, numerous wild ones such as black cherry (*Prunus serotina*) are found across North America. In California, drought-adapted hollyleaf cherry (*P. ilicifolia*) is a favorite Xerces hedgerow species. Chokecherry (*P. virginiana*), native to much of North America, has incredible tolerance for extreme cold, harsh winds, and other tough conditions: on the northern plains it is commonly planted in shelterbelts. **Note that the leaves, branches, and seeds of these and many other *Prunus* species can be toxic to livestock.**

USES

Hedgerow

Reforestation/shade garden

Pollinator nesting material or caterpillar host plant

Ornamental

Edible/herbal/medicinal

BLOOM TIME
Spring

FLOWER COLOR
Pink, white

MAXIMUM HEIGHT
80 feet (24 m)

Introduced Trees and Shrubs

NOTABLE FLOWER VISITORS

Attracts early-season bee species; the mining bee *Andrena fenningeri* specializes in pollen collection from both *Prunus* and *Salix* (willow) species. Host plant for the caterpillars of many butterflies, including eastern tiger swallowtail (*Papilio glaucus*), western tiger swallowtail (*P. rutulus*), pale swallowtail (*P. eurymedon*), coral hairstreak (*Satyrium titus*), California hairstreak (*S. californica*), striped hairstreak (*S. liparops*), Lorquin's admiral (*Limenitis lorquini*), white admiral (*L. arthemis*), Weidemeyer's admiral (*L. weidemeyerii*) and spring azure (*Celastrina ladon*).

Caterpillars of numerous large, spectacular moths also feed on these plants, including blackwaved flannel (*Lagoa crispata*), Nevada buckmoth (*Hemileuca nevadensis*), cecropia (*Hyalophora cecropia*), Cynthia (*Samia cynthia*), elm sphinx (*Ceratomia amyntor*), blinded sphinx (*Paonias excaecata*), elegant sphinx (*Sphinx perelegans*), small-eyed sphinx (*Paonias myops*), two-spotted sphinx (*Smerinthus jamaicensis*), Glover's silk (*Hyalophora columbia*), hummingbird clearwing (*Hemaris thysbe*), imperial (*Eacles imperialis*), Io (*Automeris io*), Polyphemus (*Antheraea polyphemus*), and Promethean (*Callosamia promethea*).

RIGHT Honey bee on almond blossom

SAND CHERRY

WILD PLUM

4.

Introduced Herbs and Ornamentals

Introduced herbs and ornamentals that are planted for humans are often attractive and edible for pollinators as well. Such annual and perennial plants can provide pulses of nectar and pollen for beneficial insects. Introduced herbs and ornamentals that are non-weedy can be combined with native wildflowers for many benefits: beauty, food, medicine, and spice. These are some of our favorite introduced herbs and ornamentals for pollinators and people alike.

78. BASIL

(*Ocimum* spp.)

EXPOSURE	SOIL MOISTURE
Sun	Average

A N EASY-TO-GROW ANNUAL, common sweet basil (*Ocimum basilicum*) supports bee visitors — but only if allowed to bloom. Like most kitchen herbs, it attracts a wide range of showy and beneficial insects while remaining largely pest-free itself. Inexpensive and easily available, basil is a good candidate for including in the home garden as a temporary **insectary strip**: a row of quick-flowering annuals sown between other food crops to attract beneficial insects for both pollination and pest control. At Xerces we've used basil, along with coriander and dill, to produce a simple insectary strip that blooms quickly and provides an ongoing source of herbs for the kitchen.

For maximum pollinator attractiveness, basil is most effective in massed plantings. Lone individual plants are not likely to attract much attention in the home garden when other summer-blooming plants such as sunflowers are nearby. Basil is very susceptible to frost and cold-weather injury. In cold climates, start transplants indoors for earlier-flowering plants.

USES

Ornamental

Edible/herbal/medicinal

RECOMMENDED SPECIES OR VARIETIES

H oly basil (*Ocimum tenuiflorum*) and many cultivated varieties of *O. basilicum* are readily available. Potentially better for attracting pollinators, lemon basil (*O.* × *citriodorum*) is a hybrid with typically larger, showier flowers.

NOTABLE FLOWER VISITORS

W hen planted in masses and allowed to flower, attracts bumble bees, a variety of small wild sweat bees (*Halictus* spp.), small beneficial (aphid-eating) syrphid flies, tiny beneficial wasps, and honey bees.

BLOOM TIME	FLOWER COLOR	MAXIMUM HEIGHT
Summer	White	2 feet (0.6 m)

79. BORAGE

(Borago officinalis)

ATTRACTIVE STAR-SHAPED FLOWERS turn from blue to pink on borage as they age, adding an interesting shape and texture to the garden. This annual Mediterranean herb is now occasionally cultivated as a field crop for oil (used in the cosmetics industry) and for naturopathic medicine. More commonly it's grown as a garden plant, especially in Europe, where it has historically served as a food plant, but it grows well across most of North America as well. Both flowers and leaves are edible.

For longer bloom periods, stagger the seeding dates. Borage will adapt to many soil types, but grows largest and flowers most prolifically in rich, fertile soils. The sugar in its nectar has been measured at 52% and produces a dark honey. Borage apparently secretes nectar throughout the day, and continues to yield even in cold weather.

EXPOSURE
Sun

SOIL MOISTURE
Average

USES

Ornamental

Edible/herbal/medicinal

RECOMMENDED SPECIES OR VARIETIES

In addition to the blue-flowered species, a white-flowered cultivar is available.

NOTABLE FLOWER VISITORS

While bumble bees (especially species with short tongues) visit borage, it is most important as a honey bee plant. Butterflies and other flower visitors typically ignore the flowers.

BLOOM TIME	FLOWER COLOR	MAXIMUM HEIGHT
Spring to summer	Blue	2 feet (0.6 m)

80. CATNIP
(*Nepeta* spp.)

EXPOSURE
Sun

SOIL MOISTURE
Average to dry

S EVERAL DOZEN perennial and annual species of catnip are native to Europe, Asia, and Africa. Some have escaped as naturalized weeds across the United States and Canada, while a few others are planted as garden ornamentals. Under optimal conditions most catnip species bloom for an extended period, sometimes several months. They are relatively resilient to alternating cycles of hot and cold weather as well as to dry conditions, but they grow more prolifically in fertile, damp soils. While deer avoid catnip, cats do not and can make quick work of plants that aren't protected.

The average sugar concentration reported in the nectar of catnip species ranges from 22 to 28%, and the resulting honey is dark in color and slightly spicy. While each individual catnip flower produces only a tiny amount of nectar, plants can produce a reliable honey surplus when grown in large masses.

RECOMMENDED SPECIES OR VARIETIES

A ll types of catnip are generally good bee plants, including the common true catnip (*Nepeta cataria*). The ornamental hybrid Faassen's catnip (*N. × faassenii*) is frequently planted as a ground cover or garden border plant; it grows and flowers prolifically, attracting huge numbers of honey bees and bumble bees.

USES

Ornamental

Edible/herbal/medicinal

NOTABLE FLOWER VISITORS

A ttracts honey bees and bumble bees most commonly.

BLOOM TIME
Spring to summer

FLOWER COLOR
White, blue

MAXIMUM HEIGHT
2 feet (0.6 m)

81.

CORIANDER

(Coriandrum sativum)

ALSO KNOWN AS CILANTRO, coriander is an annual herb native to Asia and North Africa. Its flowers bloom for a long period (sometimes more than a month) and produce large amounts of nectar, although it is low in sugar concentration. Coriander is often included in commercial seed mixes for beneficial insects, and organic farmers have widely adopted it for interplanting in an insectary strip among other cash crops.

EXPOSURE
Full sun to partial shade

SOIL MOISTURE
Dry to moist

RECOMMENDED SPECIES OR VARIETIES

Various cultivars have been developed, including the slow-to-flower 'Calypso' and 'Marino'. For pollinators, any low-cost, fast-flowering variety is fine.

NOTABLE FLOWER VISITORS

Honey bees and paper wasps (*Polistes* spp.) are the largest insects typically observed. A closer look will reveal large numbers of tiny dark sweat bees (*Lasioglossum* spp.), small aphid-eating syrphid flies, and many types of small beneficial wasps.

USES

Edible/herbal/medicinal

BLOOM TIME	FLOWER COLOR	MAXIMUM HEIGHT
Summer	White, pink	3 feet (1 m)

82. COSMOS
(Cosmos bipinnatus)

A POPULAR, LOW-COST summer- and fall-blooming garden annual, cosmos thrives in fairly poor soils as long as it has minimal competition, adequate soil moisture, and full sun. This member of the sunflower family is most effective at attracting pollinators when mass-planted to create large blocks of color in the landscape. Researchers in New Mexico have successfully used cosmos in combination with a few other inexpensive annuals (including dill, buckwheat, California bluebells, alyssum, and plains coreopsis) as an insectary plant with pumpkin crops, supporting predators of the squash bug and spotted cucumber beetle.

EXPOSURE
Sun

SOIL MOISTURE
Average

RECOMMENDED SPECIES OR VARIETIES

Many varieties of cosmos offer variations in color, height, petal shape, and early flowering. To attract pollinators, select simple, flat-petaled varieties in pink or white; avoid those with double petals.

NOTABLE FLOWER VISITORS

Attracts various bee species that are active in summer, including green metallic sweat bees (*Agapostemon* spp.) and long-horned bees (*Mellisodes* spp.). The latter bees tend to be specialists of the sunflower family and are most active in late summer, when cosmos is in full bloom. It is not uncommon to find male long-horned bees clustered on cosmos plants, forming overnight "bachelor parties" gathered around flowers and along plant stems. This might be a way of conserving body heat, allowing the bees to become active earlier on cool mornings.

USES

Ornamental

RIGHT Bumble bee exploring the cosmos

BLOOM TIME	FLOWER COLOR	MAXIMUM HEIGHT
Summer to autumn	White, pink, red	4 feet (1.2 m)

83. HYSSOP
(*Hyssopus officinalis*)

EXPOSURE
Full sun to partial shade

SOIL MOISTURE
Average to dry

TRADITIONALLY, HYSSOP was mass-planted as a honey plant, and in parts of the Middle East it is reportedly still an important source of honey, yielding more than 150 pounds per colony under good conditions. In North America it continues to live up to its reputation as a honey bee plant, although it is not widely planted today. True hyssop (not to be confused with wild anise hyssop, *Agastache* spp. (see page 26), is a small, aromatic, lavender-like shrub native to Southern Europe and the Middle East, where it has been used since biblical times as a kitchen herb and medicinal plant.

Hyssop is widely thought of as drought tolerant, but in our experience that seems true only on loose, well-drained soils where the plant can extend deep roots to extract water. Under good conditions this plant should reseed and spread over time to cover more ground, as long as it has space.

RECOMMENDED SPECIES OR VARIETIES

Though pink- and white-flowered forms are available, the blue-flowered is the most common, and the one with which we at Xerces have had the most success.

NOTABLE FLOWER VISITORS

Attracts honey bees and occasionally butterflies, bumble bees, and even hummingbirds in areas where it grows prolifically.

USES

Ornamental

Edible/herbal/medicinal

BLOOM TIME	FLOWER COLOR	MAXIMUM HEIGHT
Mid- to late summer	Blue, pink, white	2 feet (0.6 m)

84. LAVENDER
(*Lavandula* spp.)

IN EUROPE, LAVENDER is considered among the most important honey plants, especially in France, where its association with beekeeping is famous. Aside from the connection to honey bees, these small evergreen Mediterranean and Middle Eastern shrubs are of course equally famous for culinary and cosmetic uses. Lavender thrives in dry climates with well-drained soil (it is perfectly at home in gravel soil, for example), but it tends to suffer in cool, high-humidity climates, especially when subjected to shade. It's a good candidate for xeriscape gardens.

The sugar concentration of lavender nectar has been recorded at concentrations from 14 to 67%; the resulting honey is golden and readily granulates into small crystals, creating a smooth, butterlike texture. Beekeepers have reported 40 pounds of surplus honey per colony near commercial lavender farms.

EXPOSURE
Sun

SOIL MOISTURE
Dry

RECOMMENDED SPECIES OR VARIETIES

Several species including English lavender (*Lavandula angustifolia*) and French lavender (*L. dentata*). Various cultivated varieties and hybrids exist, including pink- and white-flowered varieties.

NOTABLE FLOWER VISITORS

All lavenders are productive bee plants, commonly attracting honey bees, mason bees (*Osmia* spp.), small carpenter bees (*Ceratina* spp.), bumble bees (*Bombus* spp.), and both native and nonnative wool carder bees (*Anthidium* spp.).

USES

Hedgerow

Ornamental

Edible/herbal/medicinal

BLOOM TIME	FLOWER COLOR	MAXIMUM HEIGHT
Summer	Purple, pink, white	3 feet (0.9 m)

85. MINT
(*Mentha* spp.)

MOST TRUE MINTS are not native but were introduced to North America, although at least one native species is widely distributed, the wild field mint (*Mentha arvensis*). Nearly all prefer rich, damp soils, such as cool stream banks. A few are less picky and will adapt even to sunny, well-drained locations. Many mints spread by underground rhizomes and can be aggressive under optimal conditions; many of the most aggressive species, however, such as the introduced peppermint (*Mentha* × *piperita*), are sterile hybrids unlikely to reseed themselves. Surplus honey production approaching 200 pounds per colony has been documented near commercial mint fields. The honey is amber in color and easily granulates, with very small crystals.

EXPOSURE
Sun to shade

SOIL MOISTURE
Average to wet

RECOMMENDED SPECIES OR VARIETIES

Of the more than 200 species of true mints (including hybrid), the two that we at Xerces are most familiar with are the common garden spearmint (*M. spicata*) and the native wild field mint (*M. arvensis*). We wish seed of this latter species were more widely available for habitat restoration projects.

NOTABLE FLOWER VISITORS

Attracts honey bees, bumble bees, many types of beneficial flies and wasps, and countless small native bees.

USES

Wetland restoration

Farm buffer/filter strip

Edible/herbal/medicinal

BLOOM TIME	FLOWER COLOR	MAXIMUM HEIGHT
Summer	White, lavender	1.5–2 feet (45–60 cm)

86. OREGANO
(Origanum spp.)

LIKE OTHER MEDITERRANEAN KITCHEN HERBS, oregano prefers sunny locations and fairly dry, alkaline soils. It is a member of the mint family, as are many other herbs. And while many oreganos are more cold-tolerant than some of their relatives (such as lavender and rosemary), some species are not truly cold-hardy and will not survive cold winters. In warm and moderate climates, however, oregano is a reliable, low-maintenance perennial. Its nectar has some of the highest documented sugar concentration of any plant, up to 76%. The honey produced from oregano is famous in parts of Greece (where it originates on remote mountain meadows), and the plant has been reported to yield surpluses of more than 40 pounds of honey per colony.

EXPOSURE
Sun

SOIL MOISTURE
Dry

RECOMMENDED SPECIES OR VARIETIES

True oregano (*Origanum vulgare*) and its various subspecies and cultivars, also sweet marjoram (*O. majorana*).

NOTABLE FLOWER VISITORS

Attracts bees, especially bumble bees and honey bees.

USES

Ornamental

Edible/herbal/ medicinal

BLOOM TIME	FLOWER COLOR	MAXIMUM HEIGHT
Summer	White, pink	2 feet (60 cm)

87. ROSEMARY

(Rosmarinus officinalis)

A COMPACT WOODY SHRUB with limited cold-hardiness, rosemary generally blooms in late winter through spring (depending on the climate), but sometimes flowers again in autumn. This wonderful evergreen plant with hemlocklike foliage and attractive blue flowers is well adapted to dry, exposed locations. It is well suited for xeriscape gardening and for use in hedges (at least in warm climates). Because rosemary will not survive harsh winters, in cold climates it is sometimes grown in containers, then moved inside during the winter months and maintained as a house plant.

Rosemary is a honey bee plant resource with sugar concentrations in the nectar of 25 to 63%. Up to 132 pounds of surplus honey per colony have been reported, including individual production rates of up to 15 pounds of rosemary honey per colony per day. The resulting honey is clear and water white.

EXPOSURE
Sun

SOIL MOISTURE
Dry

RECOMMENDED SPECIES OR VARIETIES

Cultivars have been developed with white or pink blossoms (as opposed to the common blue ones), yellow-streaked foliage, or a low, ground-hugging growth habit. Upright varieties that produce a profusion of blue flowers, such as 'Tuscan Blue', are good for bees.

NOTABLE FLOWER VISITORS

Attracts bees most commonly, especially honey bees and bumble bees. (Note that this is true with most Mediterranean kitchen herbs in the mint family that have been introduced to North America, such as lavender, oregano, and thyme.)

USES

Hedgerow

Ornamental

Edible/herbal/medicinal

BLOOM TIME	FLOWER COLOR	MAXIMUM HEIGHT
Spring	Blue	3 feet (0.9 m)

88. RUSSIAN SAGE

(Perovskia atriplicifolia)

A **VERY TOUGH NONNATIVE PERENNIAL**, covered with long-lasting spires of purple-blue blooms and silver foliage, Russian sage is ideal for difficult sites such as parking lot islands and sunny building foundations. Contrary to its common name, Russian sage is not from Russia but from south-central Asia. That geographic origin explains the plant's adaptation to extremes of heat, cold, drought, salt, and highly alkaline soils. In very cold climates, Russian sage dies back to the ground each winter. In warmer climates, its growth is more shrublike.

EXPOSURE
Sun

SOIL MOISTURE
Dry

RECOMMENDED VARIETIES

Various varieties with slightly more compact growth habits have appeared on the market. All are excellent, pest- and trouble-free bee plants.

NOTABLE FLOWER VISITORS

Attracts bumble bees, honey bees, and wool carder bees (*Anthidium* spp.). There is little documentation of the plant's honey production potential, but based on observations of honey bee visitation, Russian sage could be an important honey plant in dry climates, where large numbers are grown as ornamentals.

USES

Reclaimed industrial land/ tough sites

Ornamental

BLOOM TIME	FLOWER COLOR	MAXIMUM HEIGHT
Summer	Blue	5 feet (1.5 m)

89. THYME
(Thymus spp.)

IN GREECE, WHERE BEES FORAGE from the wild plants in their native environment, thyme and oregano are a significant and well-known source of wild honey. Thyme includes several dozen related species, most of them small, creeping evergreens that thrive in any sunny location regardless of heat or cold. In the garden they are frequently used as a ground cover, carpeting rock gardens and areas between stepping-stones. Several Xerces members use various creeping thyme species as a low-maintenance substitute for lawn grass (although the plants cannot sustain the same foot traffic as turf grass).

The average sugar concentration of thyme nectar has been recorded at 27 to 45%, with that nectar being a strong attractant. The surplus honey yields from thyme have been reported as high as 125 pounds per colony. The resulting honey is amber colored and minty smelling.

EXPOSURE	SOIL MOISTURE
Sun	Dry to moist

RECOMMENDED SPECIES OR VARIETIES

Dozens of species and cultivated varieties of thyme are available; all appear to be good bee plants.

NOTABLE FLOWER VISITORS

Attracts bumble bees and honey bees.

USES

Ornamental

Edible/herbal/medicinal

BLOOM TIME	FLOWER COLOR	MAXIMUM HEIGHT
Summer	Lavender, white	8 in. (20 cm)

Native and Nonnative Bee Pasture Plants

Native and nonnative bee pasture plants include good choices for farm, pasture, and tough sites. Such annual and perennial plants cover ground and improve soil health, while yielding lush nectar flows. Native and nonnative bee pasture plants can be combined to provide food, nests, and shelters for pollinators.

90.

ALFALFA

(Medicago sativa)

CONSIDERED THE MOST IMPORTANT honey plant west of the Missouri River, alfalfa is an important fodder and forage legume that depends on pollinators for seed production, and perennial and annual varieties support a large abundance and diversity of pollinators. The two bottom "keel" petals hold the stamen column of the alfalfa flower under tension. Visiting bees release the column and in the process are often struck in the head and dusted by pollen. Honey bees learn to bypass this process and rob nectar by probing the back of the flower with their tongues.

Plant alfalfa only in well-drained soils to reduce winterkill caused by frost-heaving of roots. Nectar flows are best following wet springs, and average sugar concentrations commonly range between 41 and 44%. Up to 300 pounds of honey per hive are reported when alfalfa fields are stocked at two hives per acre.

EXPOSURE
Sun

SOIL MOISTURE
Average

RECOMMENDED SPECIES OR VARIETIES

Many regionally adapted cultivars of alfalfa were once common, but now most breeding is conducted by large agribusinesses that emphasize hay production rather than flowering. Numerous annual and perennial varieties remain occasionally available, however. The Siberian-adapted yellow alfalfa (*M. falcata*) is one very cold-tolerant and drought-tolerant variety that is extremely useful for tough sites.

NOTABLE FLOWER VISITORS

Attracts honey bees and many beneficial insect species including leafcutter bees (*Megachile rotundata*) and alkali bees (*Nomia melanderi*), both managed as alfalfa pollinators. Host plant for caterpillars of the melissa blue (*Plebejus melissa*), marine blue (*Leptotes marina*), orange sulphur (*Colias eurytheme*), clouded sulphur (*Colias philodice*), southern dogface (*Zerene cesonia*), and eastern tailed-blue (*Cupido comyntas*) butterflies. Also hosts communities of beneficial predatory and parasitic insects and mites that are important for biological pest control.

Although alfalfa pollen has high average protein levels, it lacks the essential protein isoleucine, thus contributing to nutritional stress and colony declines in honey bees with restricted forage.

USES

Reclaimed industrial land/ tough sites

Rangeland/pasture

Farm buffer/filter strip

Cover crop

Pollinator nesting material or caterpillar host plant

BLOOM TIME	FLOWER COLOR	MAXIMUM HEIGHT
Summer	Purple, yellow	3 feet (0.9 m)

91. BUCKWHEAT
(Fagopyrum esculentum)

WELL KNOWN AS A HONEY PLANT, buckwheat produces a honey that is extremely dark and pungent — the smell is sometimes compared to that of a dead animal — with a strong flavor reminiscent of molasses. Though buckwheat is grown and eaten like a grain, it is not technically a grain but instead related to rhubarb. It is also grown as a cover crop that develops quickly in warm weather, making it a valuable green manure, soil conditioner, and weed suppressor. Buckwheat flowers prolifically during late summer, with most nectar secretion occurring in the morning. Clusters of small, shallow white flowers with pink anthers are borne at the end of multiple branched stems with heart-shaped leaves.

EXPOSURE
Sun

SOIL MOISTURE
Average

Buckwheat requires very fertile, loose, moist soil, plus cool weather for maximum nectar flow. If any one of those requirements is absent, nectar flow will be reduced by 50% or more. Honey crops may vary year to year, with yield increases up to 25 pounds per colony or 8 pounds per day for 2 to 3 weeks under favorable conditions. Average sugar concentrations are 7 to 48%. Pollen proteins at 10% are below minimum honey bee nutritional needs (20%), so other flowers mentioned in this book should be made available.

RECOMMENDED VARIETIES

Numerous regionally adapted buckwheat varieties are widely available for seed production. Consider hardiness and pest resistance when choosing varieties for seed production — although buckwheat cover crops are regularly planted without specified variety. For nectar production, avoid shatter-resistant cultivars.

USES

Cover crop

Edible/herbal/medicinal

NOTABLE FLOWER VISITORS

Attracts a variety of beneficial insects, including bees, bugs, butterflies, and wasps. Abundant buckwheat nectar also supports beneficial predatory and parasitic insects for biological pest control.

BLOOM TIME	FLOWER COLOR	MAXIMUM HEIGHT
Summer	White	4 feet (1.2 m)

92. CLOVER
(*Trifolium* spp.)

CONSIDERED THE MOST IMPORTANT group of honey plants in North America, clovers produce nectar yielding large quantities of light mild honey with enormous commercial appeal. All are important fodder plants, green manures, and cover crops, and clovers also fix nitrogen. For the greatest pollinator benefit, plant a variety of species: all are high-value bee plants. Clovers are mostly intolerant of acidic soils or drought, and dry weather can reduce nectar flow. Practically all are great pollinator plants; best location can vary among species. Many species reportedly produce more than 200 pounds of surplus honey under optimal conditions. Average nectar sugar concentration reported is 22 to 55%, and pollen protein levels are high (>25%), depending on species and location.

EXPOSURE
Sun to part shade

SOIL MOISTURE
Average

RECOMMENDED SPECIES OR VARIETIES

Numerous species and varieties of clover are cultivated, and most are prolific in nectar production, but a few are considered weedy. White Dutch clover (*Trifolium repens*) is a low-growing ground cover that tolerates mowing and is beneficial for lawns and orchards. Crimson clover (*T. incarnatum*) is a beautiful red-flowered annual often planted as a cover crop. Alsike clover (*T. hybridum*) is an excellent perennial honey plant that grows vigorously in cool climates and tolerates wetter, more acidic soils than other clover species. Red clover (*T. pratense*) is a short-lived perennial with deep nectaries that are often difficult for short-tongued bees to reach, but fine for bumble bees.

USES

Rangeland/pasture

Farm buffer/filter strip

Cover crop

Pollinator nesting material or caterpillar host plant

NOTABLE FLOWER VISITORS

Attracts a wide variety of bees, butterflies, and wasps. Host plant for caterpillars of gray hairstreak (*Strymon melinus*), greenish blue (*Plebejus saepiolus*), shasta blue (*P. shasta*), eastern tailed-blue (*Cupido comyntas*), orange sulphur (*Colias eurytheme*), clouded sulphur (*C. philodice*), Queen Alexandra's sulphur (*C. alexandra*), and southern dogface (*Zerene cesonia*) butterflies. Although important plants for beneficial insects important for biological pest control, clovers may also host tarnished plant bugs (*Lygus lineolaris*), which feed on numerous crops.

BLOOM TIME
Late spring to summer

FLOWER COLOR
White, pink, red

MAXIMUM HEIGHT
1 foot (30 cm)

93. COWPEA
(*Vigna unguiculata*)

COWPEA IS THE CATCHALL NAME for several subspecies of bean including the Chinese yard-long bean, the familiar black-eyed pea, and several others. Most of these subspecies (and their cultivars) produce exceptionally long bean pods and, depending on the variety, may be grown for human consumption, for livestock fodder, or as a green manure crop to improve soil fertility. Some shade tolerance allows cowpeas to be planted between other, taller row crops and as an understory plant in multilevel farm systems. This latter use is common in the tropics, where cowpea is farmed beneath tropical fruit trees. Cowpea is considered a valuable honey plant in some regions, such as India and parts of Africa, producing a dark, mild-flavored honey.

EXPOSURE
Full sun to partial shade

SOIL MOISTURE
Dry

RECOMMENDED SPECIES OR VARIETIES

In North America, most commercially available varieties of cowpea fall into one of two groups. Subspecies *unguiculata* includes black-eyed peas, and subspecies *sesquipedalis* includes the yard-long bean, asparagus bean, and the Chinese long bean. Of these, black-eyed peas are the most widely available and least expensive.

NOTABLE FLOWER VISITORS

The extrafloral nectaries at the base of the plant's leaf petioles feed bees as well as a variety of beneficial insects such as small wasps and syrphid flies.

USES

Cover crop

Edible/herbal/medicinal

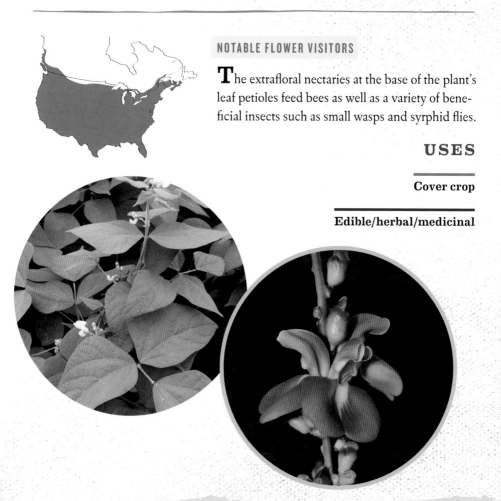

BLOOM TIME	FLOWER COLOR	MAXIMUM HEIGHT
Summer	Purple, white	3 feet (1 m)

94.
MUSTARD
(*Brassica* spp.)

VALUABLE COVER, forage, row, and specialty crops, mustards suppress pests and weeds and help to control erosion. Most species are annuals or biennials that attract a variety of bee species. In some areas many mustards are considered weedy or noxious, while some attract crop pests such as stinkbugs (many genera) and tarnished plant bugs (*Lygus lineolaris*). Mustards are sometimes planted as a low-cost cover crop for honey bees, an increasing practice near California's almond orchards. Due to high moisture requirements, they are generally low-quality honey plants with strong-flavored honey that crystallizes quickly and is difficult to extract. Reported average nectar sugar concentrations are 50 to 51%, and the pollen is high in fat (5%) and protein (25%).

EXPOSURE	SOIL MOISTURE
Sun	Average

RECOMMENDED SPECIES

Consult federal or state noxious weed information before selecting mustards for use. Field mustard or canola (*Brassica rapa*) is a winter-growing biennial oilseed, which readily attracts numerous species of bees. Chinese mustard (*B. juncea*), rapeseed (*B. napus*), and black mustard (*B. nigra*) each attract not only honey bees and native bees, but also other managed nonnative bees, such as alfalfa leafcutter bees. Related vegetable crops such as broccoli and cabbage attract bees and syrphid flies when allowed to bolt.

USES

Cover crop

Pollinator nesting material
or caterpillar host plant

Edible/herbal/medicinal

NOTABLE FLOWER VISITORS

Host plant for caterpillars of large marble (*Euchloe ausonides*), large white (*Pieris brassicae*), checkered white (*Pontia protodice*), Becker's white (*P. beckerii*), and Pacific orangetip (*Anthocharis sara*) butterflies.

BLOOM TIME	FLOWER COLOR	MAXIMUM HEIGHT
Spring to summer	Yellow	6 feet (1.8 m)

95. PARTRIDGE PEA
(Chamaecrista fasciculata)

THIS EASY-TO-GROW, heat-loving native annual legume has a promising future as a cover crop species, quickly spreading to cover the ground in a manner similar to hairy vetch. Partridge pea has nectaries at the base of its leaf petioles in addition to those in the flowers, and together these attract a huge assortment of small flies, wasps, ants, bees, and velvet ants (a kind of wingless wasp). Because partridge pea is an annual species, it does not grow well in areas already dominated by perennial vegetation. It works best on bare ground or in the early stages of prairie restoration, where it disappears after several seasons. Partridge pea grows well in burned and disturbed sites and provides food and shelter for gamebirds, songbirds, and deer throughout central, eastern, and southern North America.

EXPOSURE
Sun

SOIL MOISTURE
Average to dry

RECOMMENDED SPECIES OR VARIETIES

The USDA Natural Resources Conservation Service (NRCS) Plant Materials Centers currently list three cultivars adapted for beautification, erosion control, and habitat restoration in central plains and southern regions: 'Comanche', 'Lark Selection', and 'Riley'. Sensitive partridge pea (*Chamaecrista nictitans*) is smaller than *C. fasciculata* and has leaflets that fold when touched; it occurs in a similar range but is less widely available. Nearly one dozen other species of *Chamaecrista* have limited distributions and are native to southeastern North America. Please note that partridge pea foliage may be poisonous to cattle.

USES

Wildflower meadow/ prairie restoration

Cover crop

Pollinator nesting material or caterpillar host plant

Edible/herbal/medicinal

NOTABLE FLOWER VISITORS

Attracts leafcutter bees, bumble bees, and metallic green sweat bees. Host plant for various sulphur butterflies. Extrafloral nectaries support numerous predatory and parasitic insects that prey on numerous pest species, including ants that feed on cactus moths.

BLOOM TIME
Summer

FLOWER COLOR
Yellow

MAXIMUM HEIGHT
2 feet (0.6 m)

96.
RADISH

(Raphanus sativus)

ASIDE FROM THEIR USE as a food and fodder crop, radishes are an excellent cover crop for building soil tilth. Spring planting will typically produce flowers by summer or fall, while fall planting usually means plants bloom the following year (biennial life cycle). Radishes can often become weedy, so you may not want to grow them as a cover crop unless you can tolerate their ongoing persistence, or unless you plan to cultivate the planting area fully to remove seedlings. If you grow radishes in your vegetable garden, let a few bloom to attract beneficial insects and pollinators.

EXPOSURE
Sun

SOIL MOISTURE
Average

RECOMMENDED VARIETIES

The deep, fast-growing, large Asian daikon radishes are the most common cover crop species, frequently planted to break up soil compaction in crop fields and as a biofumigant for nematodes. Many improved oilseed radish varieties are available from seed sources. Research has shown that pollinators prefer large cultivars with yellow or pink flowers over small white-flowered radishes.

NOTABLE FLOWER VISITORS

Japanese daikon and oilseed radish varieties are especially good for attracting syrphid flies and other pollinators, including bumble bees, honey bees, mason bees, mining bees, and sweat bees.

USES

Cover crop

Edible/herbal/medicinal

BLOOM TIME
Spring, summer, or fall
(depending on planting date)

FLOWER COLOR
Yellow, pink, white

MAXIMUM HEIGHT
2 feet (0.6 m)

97. SAINFOIN
(*Onobrychis viciifolia*)

A FAMOUS HONEY PLANT in its native France, sainfoin is the source of the celebrated honey of the Gâtinais region. As recently as the 1950s this nonnative legume was an important fodder and cover crop in North America, and was more widely cultivated in some locations than alfalfa or clover. Though it resists pests and tolerates drought, it will not survive competition with weeds and so should be cultivated in well-drained, nonacidic soils, in cleaned, firm, and uniform seedbeds with sainfoin-specific *Rhizobium* inoculant. Similar to alfalfa but less hardy and

EXPOSURE
Full sun

SOIL MOISTURE
Well-drained moist soil

productive, sainfoin is preferred by foraging livestock and wildlife, and it reduces bloat and nematode parasitism in ruminant guts. The name literally means "healthy hay."

Managers of dry western rangelands are increasingly recognizing sainfoin as a valuable honey plant. The sugar concentration of its nectar is reported as high as 60%, with enhanced flows in fertile soils. Sainfoin blooms early in the season for nearly 2 weeks and attracts 10 times more bees than white clover does. The honey is yellow-white, quick to granulate, and less sweet than other honeys. Researchers have documented sainfoin honey yields of up to 350 pounds per acre.

USES

Rangeland/pasture

Cover crop

RECOMMENDED SPECIES OR VARIETIES

Seed is often expensive, but once established, perennial stands can be long lived. Many region-specific varieties are available that offer improved disease resistance, nitrogen fixation, and yields, including 'Eski', 'Melrose', 'Nova', and 'Remont'.

NOTABLE FLOWER VISITORS

Abundant nectar and pollen attract beneficial insects, including bees, butterflies, flies, and wasps.

BLOOM TIME	FLOWER COLOR	MAXIMUM HEIGHT
Late spring	Pink	2.5 feet (0.75 m)

98.
SCARLET RUNNER BEAN

(Phaseolus coccineus)

A **NATIVE OF CENTRAL AMERICA**, this vivid vining plant is commonly grown as an ornamental legume with edible starchy roots and beans. Scarlet runner bean flowers have long, tubular corollas and a diurnal pattern of flowering. Young flowers are accessible to short-tongued bees only until about midday, when they run low on nectar; after that only bees

EXPOSURE
Full sun

SOIL MOISTURE
Moist, well drained

with the longest tongues, hummingbirds, and nectar-robbers can suck the last little bit out of the flowers. With support, this vinelike plant incorporates well into tight vertical spaces, adding another dimension for pollinator gardening in small gardens. The flowers produce abundant nectar with reported sugar concentrations of 35 to 45%, thus providing high-quality bee forage.

RECOMMENDED SPECIES OR VARIETIES

Seeds of many cultivars are available commercially. 'Scarlet Runner' is the most common and productive cultivar, bearing scarlet blossoms and black-mottled maroon seeds. Other cultivars offer distinctive flower or seed colors; a few have dwarf growth habits.

NOTABLE FLOWER VISITORS

Attracts bumble bees, honey bees, and hummingbirds, with visitation and nectar robbing by carpenter bees and other short-tongued bees.

USES

Ornamental

Edible/herbal/ medicinal

BLOOM TIME
Late summer

FLOWER COLOR
Red

MAXIMUM HEIGHT
9 feet (3 m)

99.
SWEETCLOVER

(*Melilotus* spp.)

CONSIDERED ONE OF THE BEST nectar plants by many honey bee keepers, sweetclover attracts not only honey bees but also a diversity of native bees. These weedy annual and biennial legumes are very adaptable and easy to establish, even in dry and semisaline soils. Due to concerns about the invasiveness

EXPOSURE
Sun

SOIL MOISTURE
Wet to dry

of sweetclover in some regions, however, the Xerces Society does not recommend planting it in or around natural areas. For example, sweetclover is perhaps now the most invasive weed found along rivers in Alaska, where it crowds out native vegetation. In the Dakota Badlands, ecologists are concerned about how sweetclover is triggering the invasion of other nonnative plants, which more quickly establish on low-fertility soils after colonization by this nitrogen-fixing legume.

Nectar flows are best on dry soils and yield a white or greenish yellow honey flavored with hints of vanilla or cinnamon. An average of 200 pounds of surplus honey per hive is not unusual, with reported average sugar concentrations of 48 to 52%.

RECOMMENDED SPECIES OR VARIETIES

Yellow sweetclover (*Melilotus officinalis*) is the most commonly available species. For mass planting as a honey plant, however, especially in warmer climates and subtropical regions, look for fast-growing, fast-flowering annual 'Hubam', a variety of the white-flowered species *M. alba*. Overall, white sweetclover nectar flows about 2 weeks after yellow sweetclover does.

USES

Rangeland/pasture

Cover crop

Pollinator nesting material or caterpillar host plant

NOTABLE FLOWER VISITORS

Attracts not only honey bees but also a huge diversity of native bees, beneficial insects, butterflies, and wasps. Host plant for caterpillars of orange sulphur (*Colias eurytheme*), western sulphur (*C. occidentalis*), eastern tailed-blue (*Cupido comyntas*), Reakirt's blue (*Echinargus isola*), and silvery blue (*Glaucopsyche lygdamus*) butterflies.

BLOOM TIME
Late spring to summer

FLOWER COLOR
White, yellow

MAXIMUM HEIGHT
5 feet (1.5 m)

100. VETCH
(*Vicia* spp.)

DOMESTICATED NEAR THE DAWN of the agricultural revolution, vetches are grown across the Northern Hemisphere as valuable cover, fodder, and forage crops and green manures. Closely related to lentils and peas, they reduce erosion and enhance organic and no-till farming through rapid growth and nitrogen fixation. Recent research shows that vetch can also remove pollutants from soils and may be used for phytoremediation. Both native and nonnative vetches can grow aggressively and may become weedy or invasive in certain habitats and under particular growing conditions. Although occasionally mentioned as honey plants, yielding a mild white honey, some vetches have such deep flowers they are probably inaccessible to honey bees.

EXPOSURE
Sun to part shade

SOIL MOISTURE
Average

LEFT Long-horned bee exploring vetch

RECOMMENDED SPECIES OR VARIETIES

The native perennial American vetch (*Vicia americana*) and the nonnative annual hairy or winter vetch (*V. villosa*) are commonly grown as agricultural cover crops. Vetch should not be confused with the invasive crownvetch (*Securigera varia*).

NOTABLE FLOWER VISITORS

Attracts a variety of bees, especially bumble bees, honey bees, leafcutter bees, and mining bees. Host plant for caterpillars of Mexican cloudywing (*Thorybes mexicana*), funereal duskywing (*Erynnis funeralis*), western sulphur (*Colias occidentalis*), silvery blue (*Glaucopsyche lygdamus*), western tailed-blue (*Cupido amyntula*), and eastern tailed-blue (*C. comyntas*) butterflies. Vetches provide shelter for beneficial predatory insects. Annual common vetch (*Vicia sativa*) stipules have extrafloral nectaries, which support beneficial predatory or parasitic insects that prey on crop pests.

USES

Cover crop

Farm buffer/filter strip

Pollinator nesting material or caterpillar host plant

Edible/herbal/medicinal

BLOOM TIME	FLOWER COLOR	MAXIMUM HEIGHT
Spring to summer	Purple, pink	3 feet (0.9 m)

Average Number of Flower and Herb Seeds per Pound

Below is a table indicating seed volume, listed in alphabetical order by species common name.

Plant	Seeds/Pound
Alfalfa (Medicago sativa)	200,000
Aster (Symphyotrichum novae-angliae)	1,200,000
Beebalm (Monarda fistulosa)	1,400,000
Black-eyed Susan (Rudbeckia hirta)	1,700,000
Blanketflower (Gaillardia spp.)	132,000 (G. aristata); 220,000 (G. pulchella)
Blazing star (Liatris spp.)	168,000 (L. punctata); 136,000 (L. spicata)
Blue curls (Trichostema lanceolatum)	142,000
Blue vervain (Verbena hastata)	1,700,000
Buckwheat (Fagopyrum esculentum)	15,000
California poppy (Eschscholzia californica)	290,000
Clarkia (Clarkia spp.)	816,000 (C. amoena); 1,580,000 (C. unguiculata)
Clover (Trifolium spp.)	150,000 (T. incarnatum), 270,000 (T. pratense), 730,000 (T. hybridum), 750,000 (T. repens)
Coreopsis (Coreopsis spp.)	220,000 (C. lanceolata); 1,400,000 (C. tinctoria)
Coriander (Coriandrum sativum)	39,000
Cosmos (Cosmos bipinnatus)	
Cowpea (Vigna unguiculata)	4,000
Culver's root (Veronicastrum virginicum)	12,800,000
Cup plant, compass plant, rosinweed (Silphium spp.)	22,400 (S. perfoliatum); 18,400 (S. lanciniatum)
Figwort (Scrophularia marilandica)	3,000,000
Fireweed (Chamerion angustifolium)	8,000,000
Globe gilia (Gilia capitata)	680,000
Goldenrod (Solidago spp.)	775,000 (S. rigida); 1,675,000 (S. speciosa)
Gumweed (Grindelia integrifolia)	128,000
Hyssop (Hyssopus officinalis)	380,000–400,000
Hyssop, giant (Agastache spp.)	1,440,000 (A. foeniculum); 850,000 (A. urticifolia)
Ironweed (Vernonia fasciculata)	480,000
Joe-Pye weed, boneset (Eutrochium spp., Eupatorium perfoliatum)	1,520,000 (Eutrochium fistulosum)
Lobelia (Lobelia spp.)	8,000,000 (L. cardinalis); 6,400,000 (L. siphilitica)

Plant	Seeds/Pound
Lupine (*Lupinus* spp.)	22,000 (*L. perennis*); 25,000 (*L. rivularis*); 13,500 (*L. densiflorus*)
Meadowfoam (*Limnanthes alba*)	50,000
Milkweed (*Asclepias* spp.)	86,800 (*A. incarnata*); 75,000 (*A. speciosa*); 61,700 (*A. syriaca*)
Mountainmint (*Pycnanthemum* spp.)	3,200,000 (*P. virginianum*); 6,048,000 (*P. tenuifolium*)
Mustard (*Brassica rapa*)	175,000
Native thistle (*Cirsium discolor*)	102,000
Partridge pea (*Chamaecrista fasciculata*)	65,000
Penstemon (*Penstemon* spp.)	600,000 (*P. palmeri*); 400,000 (*P. digitalis*)
Phacelia (*Phacelia tanacetifolia*)	235,000
Prairie clover (*Dalea purpurea*)	275,000
Purple coneflower (*Echinacea* spp.)	85,000 (*E. pallida*); 150,000 (*E. purpurea*)
Radish (*Raphanus sativus*)	34,000
Rattlesnake master, eryngo (*Eryngium yuccifolium*)	177,000
Rocky Mountain bee plant (*Cleome serrulata*)	64,500
Sainfoin (*Onobrychis viciifolia*)	30,200
Salvia (*Salvia* spp.)	413,000 (*S. mellifera*); 149,000 (*S. azurea*)
Scarlet runner bean (*Phaseolus coccineus*)	400
Selfheal (*Prunella vulgaris*)	668,000
Sneezeweed (*Helenium autumnale*)	1,460,000
Spiderwort (*Tradescantia* spp.)	160,000 (*T. bracteata*); 144,000 (*T. occidentalis*); 128,000 (*T. ohiensis*)
Sunflower (*Helianthus* spp.)	160,000 (*H. giganteus*); 240,000 (*H. grosseserratus*); 208,000 (*H. maximiliani*); 224,000 (*H. occidentalis*)
Sweetclover (*Melilotus altissimus*)	260,000
Vetch (*Vicia* spp.)	8,000 (*Vicia sativa*), 16,000 (*V. villosa*), 40,000 (*V. americana*)
Waterleaf (*Hydrophyllum* spp.)	16,000 (*H. appendiculatum*); 4,4800 (*H. virginianum*)
Wild buckwheat (*Eriogonum* spp.)	100,000 – 200,000
Wild geranium (*Geranium* spp.)	80,000 (*G. maculatum*); 176,000 (*G. carolinianum*)
Wild indigo (*Baptisia* spp.)	27,200 (*B. alba*); 24,000 (*B. australis*); 22,400 (*B. bracteata*)
Wingstem (*Verbesina helianthoides*)	224,000
Wood mint (*Blephilia ciliata*)	6,400,000

Photo Credits

INTERIOR PHOTOGRAPHY BY © Bryan Reynolds, 18, 24R, 28, 29, 33L, 35L, 37L, 46, 47R, 69R, 75R, 79L, 87R, 89, 96, 99L, 109L, 114, 153L, 155, 159L, 180L, 185L & BR, 187 BL & R; © Jerry Pavia, 14, 24L, 26, 30, 31, 35R, 37R, 39R, 50, 54, 55, 59L, 62, 67, 70–72, 74, 75L, 77R, 79R, 80, 83L, 85, 89R, 93, 95L, 97L, 98, 102, 105L, 106, 107, 109R, 120, 123, 124, 128, 130, 140, 141, 143, 145R, 148, 149L, 150, 153R, 157, 158, 159R, 161R, 162–166, 170, 173, 175, 185TR, 186, 193, 196, 199R, 188L, 204, 205L & TR, 206, 207C & R, 208, 210–212, 213L & TR, 214, 218, 221, 224, 225, 228, 232, 233L; © Saxon Holt, 27, 32, 33R, 34, 36, 38, 40–45, 47L, 48, 56, 57R, 58, 59L, 63R, 64–66, 68, 69L, 73, 76, 77L, 78, 81, 82, 83R, 84, 86, 88, 90–92, 94, 95R, 97R, 99RT & RB, 100, 103L, 104, 105R, 108, 132, 133, 142, 144, 145L, 151, 152, 154, 156, 160, 168, 174, 180R, 188R, 190, 192, 194, 195, 198, 199L, 200–203, 205BR, 207L, 209, 213 BR, 220, 233R

WITH ADDITIONAL PHOTOGRAPHY BY © AfriPics/Alamy, 63L; © Alan Cresler/Lady Bird Johnson Wildflower Center, 127L; © Allan Munsie/Alamy, 103R; © amana images inc/Alamy, 222; © Anna Yu/Alamy, 111L; Aung/Wikimedia Commons, 137R; © bkkm/iStockphoto.com, 53R; © Blickwinkel/Alamy, 110, 219; © Bob Balestri/iStockphoto.com, 111R; © Bob Gibbons/Alamy, 52, 131R, 136, 169R; © Bruce W. Leander/Lady Bird Johnson Wildflower Center, 117R; © Cristina Lichti/Alamy, 147L; © D. Agostini Picture Library/Getty Images, 182, 183L; © DEA/E. Martini/Getty Images, 234; © DEA/S. Montanari/Getty Images, 184; © ephotocorp/Alamy, 223R; © Florapix/Alamy, 197L, 216; © Florida Images/Alamy, 138; © FLPA/Alamy, 61L; © GAP Photos/Jerry Harpur, 172; © imageBROKER/Alamy, 147R; © Inga Spence/Alamy, 237R; Javier Martin/Wikimedia Commons, 230; © John Glover/Alamy, 229R; © Ken Barber/Alamy, 149R; © Kevin Knight/Alamy, 226; © Klaus Lang/Alamy, 161L; © Leon Werdinger/Alamy, 116; © magicflute002/iStockphoto.com, 171R; Mars Vilaubi, 176; © Martin Beebee/Alamy, 169L; Michael Apel/Wikimedia Commons, 231; © Mindy Fawver/Alamy, 227; © Nigel Cattlin/Alamy, 223L; © Organica/Alamy, 60, 61R; © Patricia R. Drackett/Lady Bird Johnson Wildflower Center, 126; © Pavel Zhelev/Alamy, 119; © Pedro Pulido Grima/Alamy, 229L; © Peter Dziuk, 167L; © Peter Veilleux, East Bay Wilds, 39L, 179R; © PFMphotostock/iStockphoto.com, 134, 135; © Phil Hawkins/Getty Images, 187TL; © Philip Huston/Alamy, 178; photos courtesy of Prairie Moon Nursery, 112, 113; © Photoshot/Alamy, 57L; © Plelo/iStockphoto.com, 197R; © Richard Becker/Alamy, 53L, 237L; © RidvanArda/iStockphoto.com, 171L; © Robert Schantz/Alamy, 146; © Ron Buskirk/Getty Images, 129; © RuudMorjin/iStockphoto.com, 236; © Sally and Andy Wasowski/Lady Bird Johnson Wildflower Center, 117L, 139L, 167R; © Serge Villa/iStockphoto.com, 183R; © smilesb/iStockphoto.com, 23; © Stefan Bloodworth/Lady Bird Johnson Wildflower Center, 127R; Stan Shebs/Wikimedia Commons, 137L, 179L; Stickpen/Wikimedia Commons, 131L; © USDA-NRCS/Wikimedia Commons, 139R; © Whiteway/iStockphoto.com, 122; © Zoonar GmbH/Alamy, 118